117
Advances in Biochemical Engineering/Biotechnology

Series Editor: T. Scheper

Advances in Biochemical Engineering/Biotechnology

Series Editor: T. Scheper

Recently Published and Forthcoming Volumes

Whole Cell Sensing Systems I
Volume Editors: Belkin, S., Gu, M.B.
Vol. 117, 2010

Optical Sensor Systems in Biotechnology
Volume Editor: Rao, G.
Vol. 116, 2009

Disposable Bioreactors
Volume Editor: Eibl, R., Eibl, D.
Vol. 115, 2009

Engineering of Stem Cells
Volume Editor: Martin, U.
Vol. 114, 2009

Biotechnology in China I
From Bioreaction to Bioseparation and Bioremediation
Volume Editors: Zhong, J.J., Bai, F.-W., Zhang, W.
Vol. 113, 2009

Bioreactor Systems for Tissue Engineering
Volume Editors: Kasper, C., van Griensven, M., Poertner, R.
Vol. 112, 2008

Food Biotechnology
Volume Editors: Stahl, U., Donalies, U. E. B., Nevoigt, E.
Vol. 111, 2008

Protein – Protein Interaction
Volume Editors: Seitz, H., Werther, M.
Vol. 110, 2008

Biosensing for the 21st Century
Volume Editors: Renneberg, R., Lisdat, F.
Vol. 109, 2007

Biofuels
Volume Editor: Olsson, L.
Vol. 108, 2007

Green Gene Technology
Research in an Area of Social Conflict
Volume Editors: Fiechter, A., Sautter, C.
Vol. 107, 2007

White Biotechnology
Volume Editors: Ulber, R., Sell, D.
Vol. 105, 2007

Analytics of Protein-DNA Interactions
Volume Editor: Seitz, H.
Vol. 104, 2007

Tissue Engineering II
Basics of Tissue Engineering and Tissue Applications
Volume Editors: Lee, K., Kaplan, D.
Vol. 103, 2007

Tissue Engineering I
Scaffold Systems for Tissue Engineering
Volume Editors: Lee, K., Kaplan, D.
Vol. 102, 2006

Cell Culture Engineering
Volume Editor: Hu, W.-S.
Vol. 101, 2006

Biotechnology for the Future
Volume Editor: Nielsen, J.
Vol. 100, 2005

Gene Therapy and Gene Delivery Systems
Volume Editors: Schaffer, D.V., Zhou, W.
Vol. 99, 2005

Sterile Filtration
Volume Editor: Jornitz, M.W.
Vol. 98, 2006

Marine Biotechnology II
Volume Editors: Le Gal, Y., Ulber, R.
Vol. 97, 2005

Whole Cell Sensing Systems I

Reporter Cells and Devices

Volume Editors:
Shimshon Belkin · Man Bock Gu

With contributions by

R. Almog · P. Banerjee · S. Belkin · H. Ben-Yoav · A.K. Bhunia · L. Ceriotti · P. Colpo · R. Daniel · A. Date · S. Daunert · Tal Elad · E. Eltzov · J.T. Fleming · B. Franz · M.B. Gu · I. Karube · J.H. Lee · R.S. Marks · P. Pasini · J. Rishpon · A. Ron · E.Z. Ron · F. Rossi · A. Ruiz · Y. Shacham-Diamand · M. Shimomura-Shimizu · S. Vernick

Editors
Prof. Dr. Shimshon Belkin
The Hebrew University of Jerusalem
Institute of Life Sciences
91904 Jerusalem
Israel
shimshon@vms.huji.ac.il

Prof. Dr. Man Bock Gu
Korea University
College of Life Sciences and Biotechnolo
Amam-dong 5-ga
136-701 Seoul
Sungbuk-gu
Korea, Republic of (South Korea)
mbgu@korea.ac.kr

ISSN 0724-6145 e-ISSN 1616-8542
ISBN 978-3-642-12361-0 e-ISBN 978-3-642-12362-7
DOI 10.1007/978-3-642-12362-7
Springer Heidelberg Dordrecht London New York

Library of Congress Control Number: 2010928050

Cover design: WMXDesign GmbH, Heidelberg, Germany

Printed on acid-free paper

Springer is part of Springer Science+Business Media (www.springer.com)

Series Editor

Prof. Dr. T. Scheper

Institute of Technical Chemistry
University of Hannover
Callinstraße 3
30167 Hannover, Germany
scheper@iftc.uni-hannover.de

Volume Editors

Prof. Dr. Shimshon Belkin

The Hebrew University of
Jerusalem
Institute of Life Sciences
91904 Jerusalem
Israel
shimshon@vms.huji.ac.il

Prof. Dr. Man Bock Gu

Korea University
College of Life Sciences and Biotechnolo
Amam-dong 5-ga
136-701 Seoul
Sungbuk-gu
Korea, Republic of (South Korea)
mbgu@korea.ac.kr

Editorial Board

Advances in Biochemical Engineering/ Biotechnology Also Available Electronically

Advances in Biochemical Engineering/Biotechnology is included in Springer's eBook package *Chemistry and Materials Science*. If a library does not opt for the whole package the book series may be bought on a subscription basis. Also, all back volumes are available electronically.

For all customers who have a standing order to the print version of *Advances in Biochemical Engineering/Biotechnology*, we offer the electronic version via SpringerLink free of charge.

If you do not have access, you can still view the table of contents of each volume and the abstract of each article by going to the SpringerLink homepage, clicking on "Chemistry and Materials Science," under Subject Collection, then "Book Series," under Content Type and finally by selecting *Advances in Biochemical Bioengineering/Biotechnology*

You will find information about the

- Editorial Board
- Aims and Scope
- Instructions for Authors
- Sample Contribution

at springer.com using the search function by typing in *Advances in Biochemical Engineering/Biotechnology.*

Color figures are published in full color in the electronic version on SpringerLink.

Aims and Scope

Advances in Biochemical Engineering/Biotechnology reviews actual trends in modern biotechnology.

Its aim is to cover all aspects of this interdisciplinary technology where knowledge, methods and expertise are required for chemistry, biochemistry, microbiology, genetics, chemical engineering and computer science.

Special volumes are dedicated to selected topics which focus on new biotechnological products and new processes for their synthesis and purification. They give the state-of-the-art of a topic in a comprehensive way thus being a valuable source for the next 3-5 years. It also discusses new discoveries and applications.

In general, special volumes are edited by well known guest editors. The series editor and publisher will however always be pleased to receive suggestions and supplementary information. Manuscripts are accepted in English.

In references *Advances in Biochemical Engineering/Biotechnology* is abbreviated as *Adv. Biochem. Engin./Biotechnol.* and is cited as a journal.

Special volumes are edited by well known guest editors who invite reputed authors for the review articles in their volumes.

Impact Factor in 2008: 2.569; Section "Biotechnology and Applied Microbiology": Rank 48 of 138

Attention all Users of the "Springer Handbook of Enzymes"

Information on this handbook can be found on the internet at springeronline.com

A complete list of all enzyme entries either as an alphabetical Name Index or as the EC-Number Index is available at the above mentioned URL. You can download and print them free of charge.

A complete list of all synonyms (more than 25,000 entries) used for the enzymes is available in print form (ISBN 3-540-41830-X).

Save 15%

We recommend a standing order for the series to ensure you automatically receive all volumes and all supplements and save 15% on the list price.

Preface

The last decade has witnessed an unprecedented convergence of biological, physical, chemical, and engineering sciences that allows the construction of integrated devices that could not have been feasible earlier. Diverse combinations of biotic entities with inanimate platforms are reported that repeatedly break new grounds in the engineering of biochips, biomimetic systems, and bioarrays. One exciting front in this continuously developing field deals the deposition and immobilization of live, functioning cells onto solid surfaces for biosensor applications. The present two volumes set attempts to summarize the state of the art in this field, to highlight several specific research aspects, to describe some of the most relevant applications, and to point out what we believe are the most important future directions for whole-cell sensor systems.

To accomplish this, leading scientific authorities on biosensor-related biological, chemical, and engineering aspects have joined forces by contributing 17 comprehensive review chapters that have been divided into two "Whole-Cell Sensor Systems" volumes. Volume I addresses the two main components of such systems: the cells on the one hand and the devices on the other; the second volume is devoted to a description of a set of present and future applications of whole-cell biosensors.

We have tried to direct the manner by which these issues are addressed here to illustrate the multidisciplinary nature that is essential for such an imaginative combination of diverse scientific disciplines. It is our hope that the resulting compendium of reviews will stimulate students, teachers, and researchers from all related fields to try and tread this exciting path.

Jerusalem
Seoul

Shimshon Belkin
Man Bock Gu
Editors

Contents

Adv Biochem Engin/Biotechnol (2010) 117: 1–19
DOI: 10.1007/10_2009_18

Published online: 20 January 2010

Yeast Based Sensors

Mifumi Shimomura-Shimizu and Isao Karube

Abstract Since the first microbial cell sensor was studied by Karube et al. in 1977, many types of yeast based sensors have been developed as analytical tools. Yeasts are known as facultative anaerobes. Facultative anaerobes can survive in both aerobic and anaerobic conditions. The yeast based sensor consisted of a DO electrode and an immobilized omnivorous yeast. In yeast based sensor development, many kinds of yeast have been employed by applying their characteristics to adapt to the analyte. For example, *Trichosporon cutaneum* was used to estimate organic pollution in industrial wastewater. Yeast based sensors are suitable for online control of biochemical processes and for environmental monitoring. In this review, principles and applications of yeast based sensors are summarized.

Keywords BOD sensor • DO electrode • Environmental monitoring • Food analysis • Yeast based sensor

Contents

M. Shimomura-Shimizu and I. Karube (✉)
School of Bioscience and Biotechnology, Tokyo University of Technology, 1404-1 Katakura, Hachioji, Tokyo 1920982, Japan
e-mail: karube@bs.teu.ac.jp

Abbreviations

BOD	Biochemical (or biological) oxygen demand
DO	Dissolved oxygen
GGA	Glucose–glutamic acid
HCF	Hexacyanoferrate
HPLC	High-performance liquid chromatography
JIS	Japan Industrial Standard
LAS	Linear alkylbenzene sulfonate
RC	Redox color indicator
SP	Surface photovoltage
YM	Yeast mold

Microbes

Arxula adeninivorans, Bacillus subtilis, Candida sp., *Gluconobacter oxydans, Hansenula anomala, Issatchenkia orientalis, Kluyveromyces marxianus, Pichia methanolica, Rhodococcus erythropolis, Saccharomyces cerevisiae, Torulopsis candida, Trichosporon brassicae, Trichosporon cutaneum, Yarrowia lipolytica*

1 Introduction

Since the first microbial cell sensor was studied by Karube et al. in 1977 [1], several kinds of biosensors have been developed [2, 3]. The general developments in microbial cell sensors including yeast based sensors were introduced by many literature reports [4–14]. In yeast based sensor development, many kinds of yeast have been employed by applying their characteristics to adapt to the analyte. For example, *Trichosporon cutaneum* was used to estimate organic pollution in industrial wastewater. Yeast based sensors are suitable for online control of biochemical processes and for environmental monitoring. In this chapter, the yeast used, the principles and yeast based sensors developments, the fields of food and environmental analysis are described.

2 Yeast

The microorganisms are classified into the three kingdoms of eukarya, eubacteria, and archaea based on the sequence homology of 16S (or 18S) RNA. The eukaryote has a nuclear envelope, is not limited to unicellular organisms, and involves unicellular protozoa, yeasts, algae, and some species of molds. Eubacteria and the archaea, categorized as prokaryote, are defined as organisms having no nuclear envelope and are limited to unicellular organisms. In addition, microorganisms are also classified into three types: aerobe, facultative anaerobe, and anaerobe. Facultative anaerobes can survive in both aerobic and anaerobic conditions. Yeasts are known as facultative anaerobes. Under aerobic conditions, the microbe survives by aerobic respiration. In contrast, under anaerobic conditions, the microbe survives by anaerobic fermentation or respiration. Yeasts are chemoorganotrophs as they use organic compounds as a source of energy and do not require sunlight to grow. *Saccharomyces cerevisiae* has been used in baking and fermenting alcoholic beverages for thousands of years. The application of yeasts (*Candida* sp., *Saccharomyces* sp., *Trichosporon* sp., etc.) to the development of microbial cell sensors was reviewed recently [15].

3 Principles

In principle, the microbial cell sensors including yeast based sensors mainly involve changes in respiratory activity, which can be categorized into two groups: activation of microbial respiration by assimilation of organic compounds and inactivation of the respiration by inhibitory substances. These changes can be monitored by using a dissolved oxygen (DO) electrode. The operating principle of the yeast based sensor is to measure the change in respiration activity of immobilized yeast [16]. The yeast based sensor consisted of a DO electrode and an acetylcellulose membrane-immobilized omnivorous yeast. Further, a mediator type of microbial cell sensor has recently been developed by substitution of DO indication. The latter can be monitored directly by an electrochemical device.

3.1 DO Measuring Systems

A DO electrode is the most general transducer for the yeast based sensor. The membrane type electrode (Clark type) is widely used [17]. The Clark electrodes are classified as either galvanic electrodes or polarographic electrodes. The galvanic electrode has silver (or platinum) cathode and a lead anode, and gives rise to a potential difference. Therefore, it is a self-driven electrode and does not require an externally supplied voltage. This type of electrode is very simple; however, it has disadvantages since it shows a slower response and a shorter stability than a polarographic electrode. The polarographic electrodes consist of a platinum cathode and a silver (or silver/silver) chloride anode, both immersed in the same solution of saturated potassium chloride. A suitable polarization voltage between the anode and cathode selectively reduces oxygen at the cathode. The results of these chemical reactions are shown as a current which is proportional to the DO concentration.

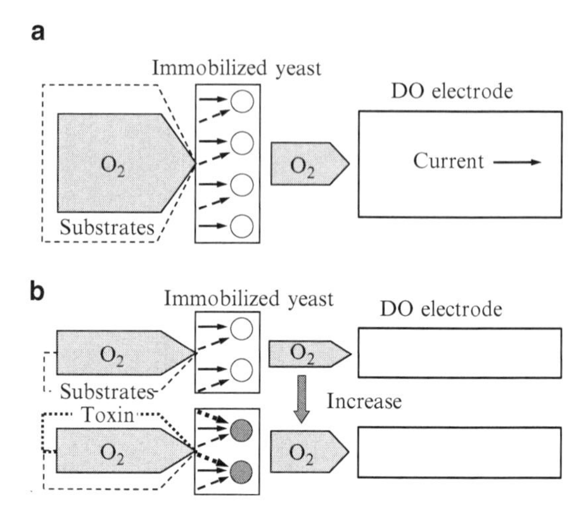

Fig. 1 Principles of yeast based sensor (DO). (**a**) Respiration activity measurement type for assimilable compounds. (**b**) Respiration activity measurement type for toxic compounds

In addition, for portable and disposable use of the yeast based sensor, miniaturized disposable type electrode (a paper-based O_2 electrode) has been developed by Yang et al. [18].

Changes due to activation of the facultative anaerobe's respiration caused by assimilation of chemical compounds (organic and/or inorganic compounds) are detected by a DO electrode [19]. These changes can be estimated as substrate concentrations. The principle in this sensor is shown in Fig. 1a. Facultative anaerobe microorganisms such as yeast are used in these sensors. When the yeast based sensor is dipped into sample solution saturated with DO, the respiratory activity of the yeast is increased, which causes a decrease in DO concentration near the membrane. Using a DO electrode, substrate concentration can be measured from the oxygen decrease.

Changes due to inactivation of the microorganism's respiration caused by toxic compounds are also detected by the DO electrode [19]. These changes can be estimated as concentrations of the toxic compound. The principle in this sensor type is shown in Fig. 1b. Yeast is used in these sensors. When the yeast based sensor is dipped into sample solution saturated with DO, the respiratory activity of the yeast is decreased, which causes an increase in DO concentration near the membrane. Using a DO electrode, the concentration of a toxic compound can be estimated from the oxygen increase.

3.2 *Electron Transfer Measuring Systems*

On the other hand, assimilation of organic compounds by microorganisms can also measure analytes using redox-active substances which can serve as electron shuttles between microorganism and electrode [19]. Electron transfers such as "mediator" or "redox color indicator (RC)" have been applied to the construction of yeast based

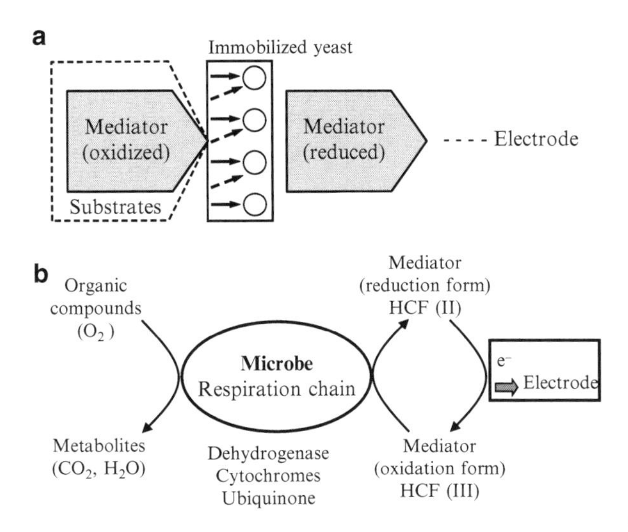

Fig. 2 Principles of mediator measurement type yeast based biosensor (**a**) and of the amperometric-mediated biosensor (**b**)

sensors. The principle of this sensor type is shown in Fig. 2a. Potassium hexacyanoferrate(III) [HCF(III)] as a mediator of oxidation form and HCF(II) as a mediator of reduction form were used for yeast based sensor development (see Fig. 2b). Generally, organic substances are oxidized by microorganisms during aerobic respiration. However, when HCF(III) is present in the reaction medium, it acts as an electron acceptor and is preferentially reduced to HCF(II) during the metabolic oxidation of organic substances. The reduced HCF(III) is then reoxidized at a working electrode (anode) which is held at a sufficiently high electric potential. Consequently, a current is generated and detected using the electrode system. The mediator type of sensor enables measurement of a certain amount of target substance without the influence of DO concentration in the analyte sample.

The sensor system utilizing electron transfer has many advantages. The solubility of mediator or color indicator is much higher than that of DO. The sensor system does not require an aeration system and can be greatly simplified to a mobile type device. The detectable potential of the mediator is low compared with that of DO. Therefore, the measurement is not influenced by reducing compounds and can be adequately performed with a normal battery, due to the fact that the electric power can be kept at the low detection potential.

3.3 Other Measuring Systems

Several other devices can also be applied to yeast based sensors [19]. The surface photovoltage (SP) technique can be applied to a yeast based sensor [20]. The SP device as a transducer is sensitive to the surface pH, ionic strength, and physical adsorption. A silicon-based SP device or a lightaddressable potentiometric sensor measures the surface potential of the device, especially the pH of the solution near the surface. The SP device can be easily fabricated by using a silicon chip. SP devices have been employed in some applications such as chemical sensors for quantification of enzyme-linked immunoassays, taste sensors, hydrogen sensors, and monitoring sensors for metabolism in mammalian cells.

A combination of microbe and the RC as an electron transfer system can be constructed for the yeast based sensor. As well as the mediator type, several advantages are expected as the features of the yeast based sensors [21].

3.4 Immobilization of Yeast

Immobilization methods of microorganisms have been studied for the yeast based sensor developments. In the case of DO electrode based sensors, gas permeability through the microorganism-immobilized membrane is significant. When it is based on the functions of living cells, a gentle condition for yeast immobilization must be required. As a porous membrane, cellulose acetate or nitrocellulose (nitrate cellulose) membrane is used for this purpose. Appropriate microorganisms culture to a cell density corresponding to an absorbance at 562 nm of the cell culture drops on to a porous cellulose acetate membrane while applying gentle suction from a water pump. The membrane containing the entrapped cells dries by air and stores at 4°C or room temperature. As a membrane, a Millipore Type HA membrane with a pore size of 0.45 μm, a diameter of 47 mm, and a thickness of 150 μm is suitable for this purpose. Both nitrocellulose and cellulose membranes can be used. The method of trapping cells between two membranes has the advantage of making the membrane with easy store and handle. Thus, as one of microbial cell sensors utilizing this microbial membrane technique, the biochemical (or biological) oxygen demand (BOD) biosensor was commercialized.

For reactor type biosensors, yeast immobilized beads acting as a support have mainly been used combined in a flow system. The controlled pore glass beads immobilize microorganisms by the adsorption method. The glass beads (20 g) with immobilized yeast (6.9×10^8 colony-forming units) can be prepared by cultivating yeast in a yeast mold (YM) medium and adding glass beads as a support. Then the yeast cells are adsorbed onto the glass beads. The glass beads were packed into a column acting as a reactor. The reactor packs glass or gel beads or a gel sheet with immobilized yeast.

4 Yeast Based Sensors Developments

4.1 Food Analysis

Yeast based sensors for food analysis and fermentation processes have been developed and the work has been reviewed in several literature reports [22–27].

Table 1 Characteristics of yeast based sensors developed in our group for food analysis

Target	Microbe	Indicator/ transducer	Measurement range ($mg\ L^{-1}$)	Ref.
(a) Ethanol (gaseous)	*T. brassicae* CBS 6382	DO/electrode	2–22.3	[28–30]
(b) Methanol (gaseous)	*T. brassicae* CBS 6382	DO/electrode	2–22.5	[28]
(c) Acetic acid (gaseous)	*T. brassicae* CBS 6382	DO/electrode	5–54	[28, 30, 36]

For fermentation process control in brewing or fuel production, monitoring of alcohol concentration is important. For monitoring of gaseous ethanol in a liquid sample, *Trichosporon brassicae* CBS 6382 [28–30] has been used as sensing element in a biosensor system which employed an acetylcellulose membrane and the DO electrode (see Table 1(a)). Gaseous methanol was also determined by *T. brassicae* CBS 6382 [28] (see Table 1(b)). In addition, several microbial cell sensors were recently developed for ethanol determination using yeast [31–33]. For koji quality control in sake brewing as the fermentation process, *S. cerevisiae* K701 and K9 were employed in the SP device [34]. The pH change due to the production of organic acids in sake brewing was determined by the SP device. Yeast activity in alcoholic fermentation was measured by the mediator system combining HCF(III) and menadione [35].

An acetic acid biosensor is required in fermentation processes and was developed using *T. brassicae* [36] (see Table 1(c)). Lactic acid was determined by a mediator type amperometric biosensor based on carbon paste electrodes modified with *S. cerevisiae* [37]. Ammonia gas or ammonium ion monitoring is not only required for food analysis and fermentation processes, but also for environmental and clinical analysis. *T. cutaneum* was immobilized in a membrane for long-term stability and was used for continuous monitoring of ammonium ion in sewage [38].

Yeast based sensors for simultaneous determinations were required for food analysis and fermentation control. Simultaneous determination of glucose, sucrose, and lactose was performed by *S. cerevisiae*, or *Kluyveromyces marxianus* [39]. Simultaneous determination of glucose and ethanol was performed by a nonselective microbial cell sensor for both glucose and ethanol using *Gluconobacter oxydans* and a glucose electrode with glucose oxidase [40] and by *G. oxydans* or *Pichia methanolica*. While the yeast cells of *P. methanolica* oxidized only ethanol, the bacterial cells of *G. oxydans* were sensitive to both substrates [41].

As with the other yeast based sensors, several applications were tried. A vitamin sensor was also studied [42]. Vitamin B_1 (thiamine) in culture broth was measured by using *S. cerevisiae* with a DO meter [42]. In this study, a possible mechanism of current generation is discussed. For the off-line determination of middle-chain alkanes, *Yarrowia lipolytica* was used [43]. For quality control of meat freshness, a yeast based sensor was applied using *T. cutaneum*, and polyamines and amino acids from meat in wash water were estimated by this sensor system [44].

4.2 Environmental Analysis

4.2.1 Organic Pollutants Monitoring Sensors

Yeast based sensors are often used for environmental analyses such as BOD measurements since they can detect or measure pollutants without the special techniques required by conventional methods [45]. BOD is one of the important pollutant indices of aquatic environments such as waste water or river because it can measure organic pollutants. Excess contamination of organic substances in aquatic environment causes serious eco-system damage. Many types of biosensors such as microbial cell sensors, enzyme sensors, and immuno sensors have been developed, and microbial cell sensors (yeast based sensors) are very suitable for environmental monitoring because of their stability [3].

BOD is conventionally measured by a 5-day BOD method (BOD_5) which is very commonly used around the world [46, 47]. Figure 3 illustrates the principle of the BOD_5. Although it is frequently used, the procedures include 5 days incubation, sample dilutions, and require special skills and laboratory facilities. The BOD_5 also includes a titration procedure which requires a number of chemicals. The data obtained by BOD_5 might be accurate, but it is not suitable for urgent and daily measurements such as monitoring of waste waters from factories. The glucose-glutamic acid (GGA) solution was defined as a standard solution for the BOD estimation of effluents. Before development of the BOD sensor, such primary effluent of wastewater or sewage from the pulp or food industry was hard to control by BOD values obtained by the BOD_5.

To overcome these disadvantages, a biosensor for BOD estimations was first developed in 1977 [48, 49]. The biosensor was also the first whole-cell biosensor. Biosensors using whole-cells of microorganisms are called microbial cell sensors, which exploit metabolic functions of living cells. The BOD_5 uses a consortium of microorganisms, but one of the microorganism species was used in this sensor to obtain high reproducibility of BOD estimation values. Since whole-cell biosensors or microbial cell sensors have advantages such as low cost, long life time, and low environmental impact, they have frequently been applied to environmental monitoring [50].

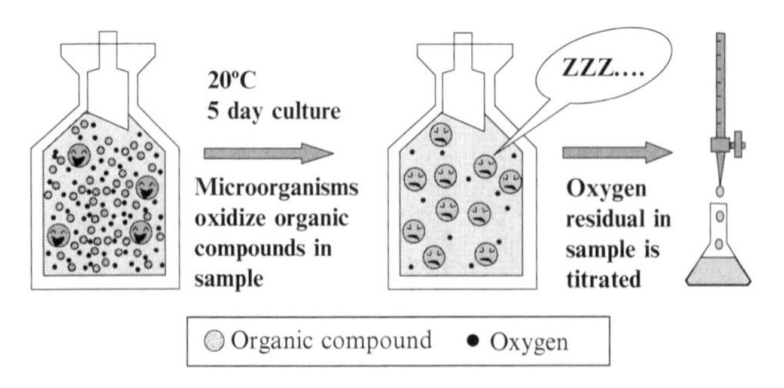

Fig. 3 Conceptual diagram in the principles of BOD_5

4.2.2 BOD_{DO} Sensors

With such a background, the necessity of the BOD monitoring was pointed out. BOD_{DO} sensor consisted of a DO electrode and immobilized omnivorous yeast (see Sect. 3.1). Figure 4 illustrates the principle of the BOD_{DO} sensor. Immobilized yeast can oxidatively degrade most organic compounds (pollutants) in waste water samples with high respiration activity. Yeast consumes DO in samples when it oxidatively degrades organic compounds, and the DO electrode measures the reduction of the DO as current decrease (sensor response). Thus, we examined the use of a single strain and fabrication of a continuous flow system for automatic BOD estimation [16] (see Table 2(Ia)). *T. cutaneum* AJ 4816 (IFO 10466) was employed as a single strain for a BOD_{DO} sensor. This sensor was able to estimate BOD in 20 min without special skills. The correlation of the sensor response and the BOD values was linear between 10 and 40 mg L^{-1} when GGA standard solution

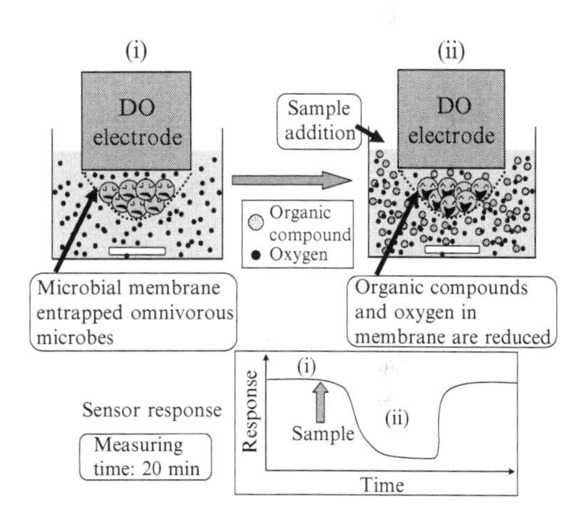

Fig. 4 Conceptual diagram in the principles of BOD_{DO} sensor

Table 2 Characteristics of yeast based sensors developed in our group for environmental analysis

Target	Microbe	Indicator/transducer	Measurement range (mg L^{-1})	Ref.
(Ia) BOD	*T. cutaneum* IFO 10466 (AJ 4816)	DO/electrode	10–40	[16]
(Ib) BOD	*T. cutaneum* IFO 10466	DO/chip electrode	1.0–18 (det.lim. 0.2)	[53]
(Ic) BOD	*T. cutaneum* IFO 10466	DO/chip electrode (five electrode array)	8.0–32	[55]
(Id) BOD	*T. cutaneum*	DO/electrode	2.0–20	[64]
(Ie) BOD	*T. cutaneum* IFO 10466 (AJ 4816)	pH/SP	10–100	[20]
(II) LAS	Strain A and *T. cutaneum*	DO/electrode	0–4	[69]
(IIIa) CN^-	*S. cerevisiae* IFO 0337	DO/electrode	0.008–4	[71]
(IIIb) CN^-	*S. cerevisiae*	DO/electrode	0.004–0.4	[72]
(IIIc) CN^-	*S. cerevisiae*	Two DO/electrode	0–0.4	[73]

was measured. The BOD_{DO} sensor was able to measure several kinds of untreated waste water from fermentation plants [16].

Comparison of biodegradation characteristics of organic compounds between the BOD_{DO} sensor method using *T. cutaneum* AJ 4816 [16] was performed and compared with the conventional BOD_5 method [51]. The sensor showed low BOD values compared with the BOD_5 method when soluble starch and lactose were employed for experiments. This might be caused from the slow decomposition rate of these compounds by the immobilized yeasts. On the other hand, the sensor showed high BOD values compared with the BOD_5 method when ethyl alcohol and acetic acid were employed. These results suggested the oxidation rate of ethyl alcohol and acetate to be faster than that of some standard substrates such as GGA. As a new method for the BOD determination of primary effluent, a sensor method was established which was defined as a Japanese Industrial Standard (JIS) (JIS K 3602) in 1990 [52].

Yang et al. developed disposable DO electrode chips that can be applied to the BOD estimation for field monitoring (see Table 2(Ib)). At first, the single DO electrode ($15 \times 2 \times 0.4$ mm) was constructed on silicon substrates using micro-machining techniques [53]. This electrode is of the Clark type and *T. cutaneum* was directly immobilized on the electrode surface using an ultraviolet cross-linking resin. For measuring the GGA standard solutions, potential of -1021 mV was applied to the working electrode. From the results, this DO electrode chip was shown to enable measurements between 1.0 and 18 mg O_2 L^{-1} BOD. For BOD estimation using the DO electrode chip, a dynamic transient measuring method was adopted and compared with the steady-state measuring BOD_5 method [54]. In the study, the measuring time was dramatically reduced.

Subsequently, using thin film technology, they fabricated an array type DO electrode [55] (see Table 2(Ic)). One array chip was formed with five DO electrodes (see Fig. 5). Each DO electrode on the array comprised an Ag/AgCl anode and a silver cathode. Electrolyte, containing polyvinylpyrrolidone, soaked both anode and cathode in each DO electrode before use. The DO electrode surface consists of

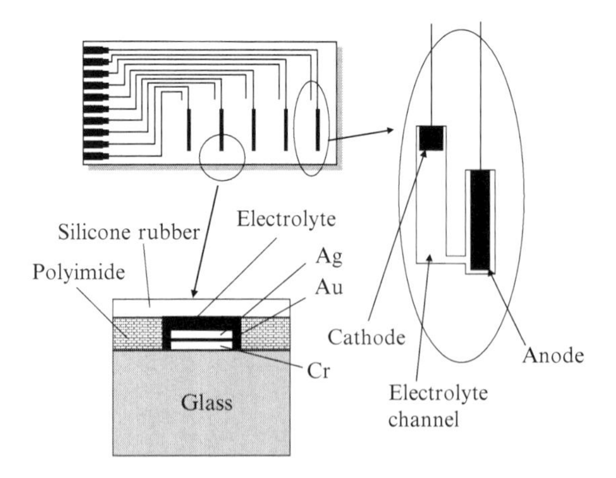

Fig. 5 Schematic diagram of an array type DO electrode

a gas permeable membrane. To apply to BOD determination, *T. cutaneum* was immobilized onto the cathode using the photo-crosslinkable resin. Using artificial and domestic samples, the BOD values were compared with the BOD_5 and this sensor method and the results agreed well with those determined using both methods.

Other BOD_{DO} sensors were also developed using different kinds of microbes, for example *Arxula adeninivorans* (salt-tolerant yeast) [56–58], *Hansenula anomala* [59], *Torulopsis candida* [60], and combinations of two microbes such as *Bacillus subtilis* and *T. cutaneum* [61], *Issatchenkia orientalis* and *Rhodococcus erythropolis* [62].

4.2.3 Commercial BOD_{DO} Sensors

About 800 instruments involving BOD_{DO} sensors have been commercialized to date (Central Kagaku Co., Tokyo, Autoteam GmbH, Berlin, etc.) since the first commercial BOD_{DO} sensor was produced by Nisshin Electric Co. Ltd in 1983 [45]. Several companies in Japan and other countries tried to commercialize the BOD_{DO} sensor. These sensors are based on the first BOD_{DO} sensor which detects the change in the respiration activity of the immobilized microorganisms by a DO. These commercialized BOD_{DO} sensors have been described in several reviews [3, 45, 50, 63].

The BOD_{DO} sensor method was established as JIS in 1990 (JIS K3602) [52]. Central Kagaku Co. offers several types of commercial BOD_{DO} sensors according to JIS K3602 in cooperation with Nisshin Electric Co. Ltd which manufactures the BOD_{DO} sensors. Figure 6 shows a photo of the deferred type BOD_{DO} sensor. This is available from Central Kagaku Co. [64]. This sensor "BOD 3300" is used for in situ continuous monitoring of samples such as waste waters. The sensor can measure up to BOD 500 mg L^{-1} in 30–60 min.

Their latest BOD_{DO} sensor is a bench top instrument "QuickBOD α-1", the detection limit of which is BOD 2 mg L^{-1}, only requiring GGA standard solution and phosphate buffer for the daily measurements [64] (see Table 2(Id)). Figure 7 shows a photo of the desktop type BOD_{DO} sensor (Central Kagaku Co.). The sandwich method using two porous membranes was used for immobilization of *T. cutaneum* to obtain high reproducibility and stability. In this system, a buffer solution as a carrier solution is continuously passed through the inner system. Antibacterial tubes are used in the systems to maintain clear conditions. When a sample solution is introduced into the sensor inside by a peristaltic pump, the sample put into contact with the surface of a microbial electrode after aeration by an air pump. Here, aeration of the carrier and sample solutions are required for the saturation of DO concentration. Then the signal obtained from the microbial electrode is sent to a microprocessor, and the sample solution is moved to a waste water bottle. A detergent solution for washing the inner system and standard solution for the calibration are used.

Fig. 6 Photographs of BOD sensors of a deferred type (Central Kagaku Co.)

4.2.4 Other BOD Sensors

According to Murakami et al., the SP technique was applied to a BOD sensor using *T. cutaneum* as a biosensing element [20] (see Table 2(Ie)). The construction of an SP device for BOD determination is described as follows. One side of the silicon chip in the SP device has an insulating layer of silicon nitride as the sensor side. The layer was deposited by chemical vapor deposition on a thermally oxidized layer. The other side (back side) has two regions. One region consists of a deposited metal layer for ohmic contact as a sensor window of 1×1 mm and the other region is not coated and remains transparent. The microbial membrane, which was prepared by sandwiching *T. cutaneum* (1 mg cell wet weight) between two filters made of

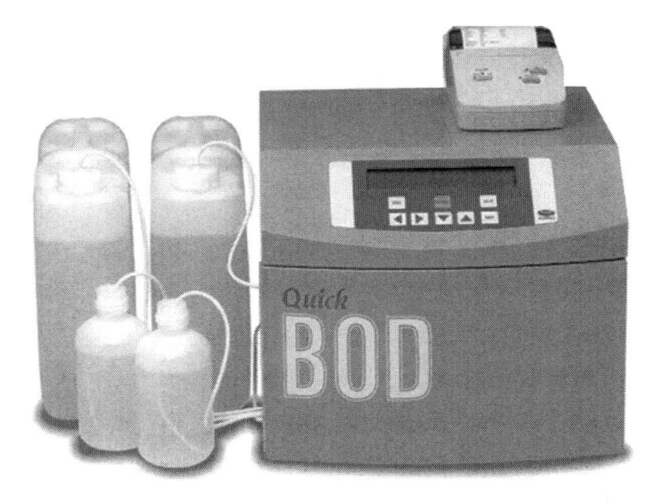

Fig. 7 Photographs of BOD sensors of desktop type (Central Kagaku Co.)

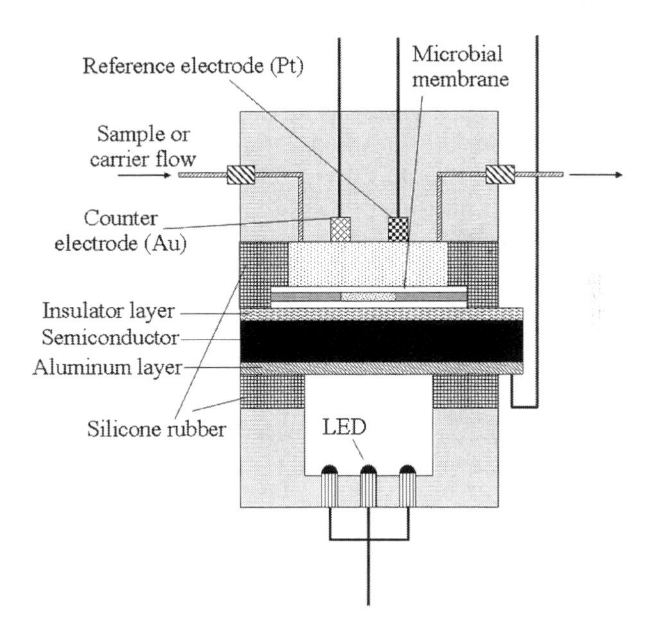

Fig. 8 Schematic diagram of a flow cell in a SP yeast based sensor system

acetylcellulose membranes, was put onto the SP device and held between the device and a silicone sheet to give a flow cell (see Fig. 8) [20]. Illumination from the back side of the SP device induces a photovoltage.

When the sensor side of the SP devices is immersed in sample solution and the potential of the solution is biased against the bulk silicon of the device using the ohmic contact, the silicon/solution system acts as metal insulator semiconductor structure. Then frequently modulated illumination induces an alternating photocurrent. This alternating photocurrent was treated at an SP controller and a personal computer. A carrier solution containing 1 mM phosphate buffer (pH 7.0) with 0.15 M NaCl was pumped at a flow rate of 250 mL min^{-1} until the photocurrent became stable. An analyte solution containing organic compounds was injected into the flow cell by a peristaltic pump. When *T. cutaneum* in the membrane takes organic compounds as substrates, pH will be changed by production of acidic substances such as the carbonate ion and organic acid as results of metabolism occurred in the cell of *T. cutaneum*. The changes of pH were sensitively detected by the SP devices. The results in the BOD estimation of wastewaters obtained by both SP-based sensor and conventional BOD_5 were well correlated between 0 and about 180 mg O^2 L^{-1} BOD [20].

Using the eukaryote *S. cerevisiae*, a double mediator system combining HCF (III) and menadione was studied [65, 66]. Baronian et al. revealed that menadione (vitamin K3; lipophilic mediator) can penetrate the outer cell membrane. Roustan et al. applied the double mediator system for measuring yeast activity in alcoholic fermentation [35]. Heiskanen et al. compared menadione and menadione bisulfite using yeast, and they revealed that hydrophobic menadione was superior to its water-soluble bisulfite derivative for probing living cells [67]. Yeasts are easily handled, omnivorous to many kinds of organic substances, and stable even in saline solutions. Thus, we have applied the double mediator system to BOD estimation employing baker's yeast *S. cerevisiae* [68].

Recently we have developed a highly sensitive and reproducible RC type BOD sensor using baker's yeast and a temperature-controlling system providing a three-consecutive-stir unit [21]. When the incubation mixture was incubated for only 10 min at 30°C, a calibration curve for GGA concentration was obtained between 1.1 and 22 mg O_2 L^{-1} ($r = 0.988$, six points, $n = 3$). The reproducibility of the optical responses in the calibration curve was 1.77% (average of relative standard deviations). This method was superior to the available BOD_{DO} sensor (Central Kagaku Co.) in the detection limit (available BOD sensor's value, 2 mg O_2 L^{-1}), dynamic range (2–20 mg O_2 L^{-1}), reproducibility (5%), and measuring interval (30 min).

4.2.5 LAS Sensors

Numerous yeast based sensors have been fabricated since the development of BOD sensor as the first microbial cell sensor. Linear alkylbenzene sulfonates (LASs) are most commonly used for the production of detergents in synthetic anionic surfactants. Biodegradation of LSA by microorganisms requires several days. However, LAS has toxicity in itself and also contributes to increase the toxicity of other pollutants in the aquatic environment. A large amount of LASs contained in domestic wastewater are allowed to flow into rivers and streams.

In Japan, several ppm of anionic surfactants in a polluted river was reported. As conventional methods for determining LAS, the methylene blue active substances method and high performance liquid chromatography (HPLC) are commonly employed. However, these methods are time consuming and complex and require large amounts of organic reagents.

For the simple and rapid determination of LAS concentration, detergent biosensors indicating LAS were developed using LAS degrading bacteria (strain A) and *T. cutaneum* for river water monitoring [69]. This is a reactor type sensor system consisting of immobilized microbes and a DO electrode. The microbes were immobilized in calcium alginate beads and the beads were packed into two columns (reactors). In this study, correlation of LAS concentration as measured by the sensor (strain A) and a conventional HPLC system was firstly obtained by using several river waters containing LAS. However, the results showed low value of the correlation coefficient ($r = 0.57$). This result showed that the river waters were contained many and varied substances and the substances could affect the sensor responses.

To overcome this problem, another sensor system (sensor B) using yeast, *T. cutaneum* which is an omnivorous microbe was employed to deduct a response of the strain A sensor (sensor A) from a response of the *T. cutaneum* sensor as the correction sensor (see Table 2(II)) [69]. As a result, good correlation between the current decrease measured by sensors A and B was obtained ($r = 0.93$). Using this correction method, correlation of LAS concentration as measured by this sensor method and the conventional HPLC method was compared and the correlation obtained by this examination was dramatically improved ($r = 0.98$). In addition, this sensor method was not influenced by the NaCl in the river sample – at least 17.3 g L^{-1} chloride ion (Cl^-). Thus, to prevent the influence of coexisting substances dissolved in a real sample, a dual sensing system was constructed. Finally, LAS monitoring of river water by this sensor method was performed and changes of LAS concentration in river water mostly corresponded with human life cycle in a day. The optimized LAS sensor could measure 0.2 mg L^{-1} of LAS in 15 min and the sensor was tested for continuous river water monitoring in situ. Thus, a practical biosensor was constructed for water quality monitoring of river waters polluted by synthetic anionic surfactants [45].

4.2.6 Cyanide Sensors

Cyanide compounds are highly toxic to fishes or animals including humans [70]. Nevertheless, cyanide is widely utilized in industrial applications, especially for electroplating. Occasional accidents have taken place when industrial plants discharge cyanide into environmental water. The lethal dose is in the range of 0.5–3.5 mg kg^{-1} body weight. Therefore, to regulate the discharge of cyanide into the environment, the Water Pollution Control Law in Japan stipulates 1 mg L^{-1} of cyanide (38.5 μM) as the maximum concentration of cyanide allowed in wastewater. Several cyanide biosensors have been developed following this principle [71–73].

The first cyanide sensor was made by employing a DO electrode entrapped in yeast with an oxygen-permeable membrane, which was confirmed as a cyanide sensor (see Table 2(IIIa)) [71]. The yeast *S. cerevisiae* IFO 0337 was selected as a sensitive microorganism to cyanide. Yeast was cultivated in a YM medium at pH 5.8, harvested by centrifugation, and then washed in a Tris-HCl buffer. The yeast was entrapped between two porous cellulose nitrate membranes. The microbial membrane incorporating the yeast was placed on the Teflon membrane cover of a Clark-type DO electrode and covered with a nylon mesh as a protective layer. The yeast in the microbial membrane can take up glucose, oxygen, and other nutritious substances through the porous membrane, being exposed to cyanide as well when it exists in the solution. This sensor was able to determine the cyanide ion (CN^-) with a linear range between 8.0 and 4000 μg L^{-1} (0.3–150 μM), which indicates that this cyanide biosensor employing a reactor and a flow system has possible applications for the monitoring of cyanide.

The second sensor employed a flow system (see Table 2(IIIb)) [72]. Yeast was immobilized on glass beads acting as support (see Sect. 3.4); then the beads were packed into a column acting as a reactor. The system employed two electrodes, and the reactor was placed between them. A buffer solution containing 150 mg L^{-1} glucose was saturated with oxygen and carried by using a peristaltic pump. When the cyanide solution was passed through the reactor, the respiration of yeast immobilized on glass beads was inhibited, and the amount of oxygen consumed by yeast decreased. As a result, the difference between postelectrode and pre-electrode current output was reduced and considered as a response to cyanide. This system was able to detect the cyanide ion with a linear range between 0 and 400 μg L^{-1} (0–15 μM), under the conditions of a flow rate of 4.5 mL min^{-1} and 25°C. These results indicate the possible use of this sensor in the construction of a flow sensor system that can be applied to continuous monitoring for preventing the discharge of cyanide from wastewater.

To ascertain applicability, the third sensor was developed for determining cyanide in river water using an improved previous sensor system (see Table 2 (IIIc)) [73]. Then Chitopearl™ beads were chosen as the most sensitive support to cyanide. The reactor used had a 22-mm diameter, a cross-sectional area of 381 mm^2, and a length of 90 mm. The temperature was not very effective to sensor responses from 7 to 29 C. Thus, the sensor employed a double electrode system and the reactor was set up between the two electrodes. Under optimized conditions consisting of a flow rate of 4.5 mL min^{-1}, a 10 mM Tris-HCl solution (pH 8.0) containing glucose of 150 mg L^{-1}, and ambient temperature (approximately 20 C), the sensor responded to CN^- at a linear range from 0 to 400 μg L^{-1} (0–15 μM). As a result, the sensor was sensitive enough to detect cyanide contamination from industrial plants in river water.

Next, in order to determine the application of the sensor to river water, distilled water (pH 5.7) was used instead of the buffer solution as carrier solution and the distilled water did not affect the sensor response to cyanide [73]. The concentration of sodium chloride, 10 mg L^{-1}, did not affect the sensor response. In addition, the maximum concentration of Mn^{2+}, Fe^{2+}, and Zn^{2+} in the Tama River in Japan did not have a significant effect on the sensor response. Six river water samples were used for cyanide detection, all of them showing a linear response between 0 and 400 μg L^{-1} (0–15 μM) CN^-. Similar responses to cyanide were obtained in spite of differences in Cl^- concentrations, BOD values, and pH of river water samples. Herbicides such as simazine can affect the sensor response by respiratory inhibition of *S. cerevisiae*. This cyanide sensor thus proved its suitability for monitoring river water. The sensor response was maintained for 9 days but was lost after that.

Biosensors employing inhibition of microbial respiration are also affected by other toxic compounds, such as pesticides and herbicides. Therefore, this kind of sensor can be used for estimating total toxicity around a polluted water area. Recently, along with other developments, luminescence-based genetically engineered microbial cell sensors for water toxicity monitoring were developed using *S. cerevisiae* [74]. Green fluorescent protein-based genetically engineered microbial cell sensors were developed using *S. cerevisiae* for genotoxicity monitoring [75].

5 Outlook

In this review of yeast based sensors, many kinds of developments have been introduced for environmental monitoring and food analysis [3, 12]. Especially in the analysis of environmental samples, it is necessary to perform rapid, simple, and multiple analyses because it highly desirable to determine the results quickly as well as to analyze numerous samples simultaneously. Some techniques developed for BOD sensors were useful as were some other yeast based sensors. In most yeast based sensors, selectivity is a shortcoming. Thus, as molecular recognition elements, it has mainly been genetically transformed microbes that have recently been engineered for selective measurements [76]. For example, BOD sensing yeast, *T. cutaneum* IFO 10466, was genetically transformed using a plasmid, pAN 7-1, for luminous BOD sensing [77]. By progression of this yeast based sensor technique, the application will be extended, especially for environmental monitoring and food analysis fields as described before.

References

1. Karube I, Mitsuda S, Matsunaga T, Suzuki S (1977) J Ferment Technol 55:243
2. Nakamura H, Karube I (2003) Anal Bioanal Chem 377:446
3. Nakamura H, Shimomura-Shimizu M, Karube I (2008) Adv Biochem Engin/Biotechnol 109:351
4. Wilson K, Goulding KH (1986) A biologist's guide to principles and techniques of practical biochemistry, 3rd edn. Edward Arnold, London
5. Karube I (1987) Micro-organisms based sensor. In: Turner APF, Karube I, Wilson GS (eds) Biosensors: fundamentals and application. Oxford University Press, New York

6. Karube I, Suzuki M (1990) Microbial biosensors. In: Cass AEG (ed) Biosensors: practical approach. IRL, Oxford
7. Kubo I, Sode K, Karube I (1991) Whole-organism based biosensors and microbiosensors, vol 1. In: Turner APF (ed) Advances in biosensors. JAI, London
8. Karube I, Chang MES (1991) Microbial biosensors. In: Blum LJ, Coulet PR (eds) Handbook of biosensors. Marcel Dekker, New York
9. Karube I, Nakanishi K (1994) Curr Opin Biotechnol 5:54
10. Karube I, Nomura Y (1996) Biosensors using immobilized living cells. In: Willaert RG, Baron GV, De Backer L (eds) Immobilized living cell systems. Wiley, New York
11. Bousse L (1996) Sens Actuators A 34:270
12. Nomura Y, Karube I (2000) Biosensors, vol 1. In: Lederberg J, Alexander M, Bloom BR, Hopwood D, Hull R, Iglewski BH, Laskin AI, Oliver SG, Schaechter M, Summers WC (eds) Encyclopedia of microbiology, vol. 1, 2nd edn. Academic, Tokyo
13. D'Souza SF (2001) Biosens Bioelectron 16:337
14. Lei Y, Chen W, Mulchandani A (2006) Anal Chim Acta 568:200
15. Baronian KHR (2004) Biosens Bioelectron 19:953
16. Hikuma M, Suzuki H, Yasuda T, Karube I, Suzuki S (1979) Eur J Appl Microbiol Biotechnol 8:289
17. Clark LC Jr, Wolf R, Granger D, Taylor Z (1953) J Appl Physiol 6:189
18. Yang Z, Suzuki H, Sasaki S, Karube I (1997) Anal Lett 30:1797
19. Nakamura H, Karube I (2006) Microbial biosensors. In: Grimes CA, Dickey EC, Pishko MV (eds) Encylopedia of sensors, vol 6. American Scientific Publishers, California
20. Murakami Y, Kikuchi T, Yamaura A, Sakaguchi T, Yokoyama K, Tamiya E (1998) Sens Actuators B 53:163
21. Nakamura H, Kobayashi S, Hirata Y, Suzuki K, Mogi Y, Karube I (2007) Anal Biochem 369:168
22. Karube I, Suzuki S (1983) Application of biosensor in fermentation processes. In: Tsao GT (ed) Annual reports on fermentation processes, vol 6. Academic, New York
23. Karube I (1985) Biosensors in fermentation and environmental control. In: Arbor BA (ed) Handbook of biotechnology. Science Publishing, New Jersey
24. Karube I, Tamiya E (1987) Food Biotechnol 1:147
25. Karube I, Sode K (1989) Biosensors for fermentation process control. In: Ghose TK (ed) Bioprocess engineering: the first generation. Chapman Hall, London
26. Ohashi E, Karube I (1993) Food Control 4:183
27. Karube I (1994) Application of biosensors in measurement of food. In: Karube I (ed) On-line sensors for food processing. Gordon and Breach, Yverdon
28. Karube I, Suzuki S, Okada T, Hikuma M (1980) Biochimie 62:567
29. Hikuma M, Kubo T, Yasuda T, Karube I (1979) Biotechnol Bioeng 21:1845
30. Karube I, Suzuki S (1980) Enzyme Eng 5:263
31. Tkac J, Vostiar I, Gorton L, Gemeiner P, Sturdik E (2003) Biosens Bioelectron 18:1125
32. Rotariu L, Bala C, Magearu V (2004) Anal Chim Acta 513:119
33. Liao MH, Guo JC, Chen WC (2006) J Magn Magn Mater 304:e421
34. Chiyo T, Matsui K, Murakami Y, Yokoyama K, Tamiya E (2001) Biosens Bioelectron 16:1021
35. Roustan JL, Sablayrolles JM (2003) J Biosci Bioeng 96:434
36. Hikuma M, Kubo T, Yasuda T, Karube I, Suzuki S (1979) Anal Chim Acta 109:33
37. Garjonytea R, Melvydasb V, Malinauskas A (2006) Bioelectrochemistry 68:191
38. Ikeda M, Hachiya H, Ito S, Asano Y, Imato T (1998) Biosens Bioelectron 13:531
39. Svitel J, Curilla O, Tkac J (1998) Biotechnol Appl Biochem 27:153
40. Reshetilov AN, Lobanov AV, Morozova NO, Gordon SH, Greene RV, Leathers TD (1998) Biosens Bioelectron 13:787
41. Lobanov AV, Borisov IA, Gordon SH, Greene RV, Leathers TD, Reshetilov AN (2001) Biosens Bioelectron 16:1001

42. Akyilmaz E, Yasa I, Dinckaya E (2006) Anal Biochem 354:78
43. Alkasrawi M, Nandakumar R, Margesin R, Schinner F, Mattiasson B (1999) Biosens Bioelectron 14:723
44. Yano Y, Numata M, Hachiya H, Ito S, Masadome T, Ohkubo S, Asano Y, Imato T (2001) Talanta 54:255
45. Nomura Y, Shimomura-shimizu M, Karube I (2007) Environmental biochemical oxygen demand and related measurement. In: Cullen D, Lowe C, Marks R, Weetall HH, Karube I (eds) Handbook of biosensors and biochips. Wiley, England
46. Japanese Industrial Standard Committee (1993) Testing methods for industrial waste water JIS K 0102. JIS, Tokyo
47. American Public Health Association (1986) Standard methods for examination of water and wastewater. American Public Health Association, Washington
48. Karube I, Matsunaga T, Suzuki S (1977) J Solid-Phase Biochem 2:97
49. Karube I, Matsunaga T, Mitsuda S, Suzuki S (1977) Biotechnol Bioeng 19:1535
50. Karube I, Nomura Y, Arikawa Y (1995) Trends Anal Chem 14:295
51. Bond RG, Straub CP (eds) (1973) Handbook of environmental control. CRC, Cleveland
52. Japanese Industrial Standard Committee (1993) Testing methods for industrial waste water JIS K 3602. JIS, Tokyo
53. Yang Z, Suzuki H, Sasaki S, Karube I (1996) Appl Microbiol Biotechnol 46:10
54. Yang Z, Suzuki H, Sasaki S, McNiven S, Karube I (1997) Sens Actuators B 45:217
55. Yang Z, Sasaki S, Karube I, Suzuki H (1997) Anal Chim Acta 357:41
56. Tanaka H, Nakamura E, Minamiyama Y, Toyoda T (1994) Water Sci Technol 30:215
57. Lehmann M, Chan C, Lo A, Lung M, Tag K, Kunze G, Riedel K, Gruendig B, Renneberg R (1999) Biosens Bioelectron 14:295
58. Chan C, Lehmann M, Tag K, Lung M, Gotthard K, Riedel K, Gruendig B, Renneberg R (1999) Biosens Bioelectron 14:131
59. Riedel K, Renneberg R, Kuhn M, Scheller F (1988) Appl Microbiol Biotechnol 28:316
60. Kim MN, Kwon HS (1999) Biosens Bioelectron 14:1
61. Jia J, Tang M, Chen X, Qi L, Dong S (2003) Biosens Bioelectron 18:1023
62. Heim S, Schnieder I, Binz D, Vogel A, Bilitewski U (1999) Biosens Bioelectron 14:187
63. Nomura Y, Chee GJ, Karube I (1998) Field Anal Chem Technol 2:333
64. Central Kagaku Co. http://www.aqua-ckc.jp/ (Sep. 2009)
65. Yashiki Y, Yamashoji S (1996) J Ferment Bioeng 82:319
66. Baronian KHR, Downard AJ, Lowen PK, Pasco N (2002) Appl Microbiol Biotechnol 60:108
67. Heiskanen A, Yakovleva J, Spegel C, Taboryski R, Koudelka-Hep M, Emneus J, Ruzgas T (2004) Electrochem Commun 6:219
68. Nakamura H, Suzuki K, Ishikuro H, Kinoshita S, Koizumi R, Okuma S, Gotoh M, Karube I (2007) Talanta 72:210
69. Nomura Y, Ikebukuro K, Yokoyama K, Takeuchi T, Arikawa Y, Ohno S, Karube I (1998) Biosens Bioelectron 13:1047
70. Ikebukuro K, Nakamura H, Karube I (1999) Cyanides. In: Nollet LML (ed) Handbook of water analysis. Marcel Dekker, New York, p 367
71. Nakanishi K, Ikebukuro K, Karube I (1996) Appl Biochem Biotechnol 60:97
72. Ikebukuro K, Honda M, Nakanishi K, Nomura Y, Masuda Y, Yokoyama K, Yamauchi Y, Karube I (1996) Electroanalysis 8:876
73. Ikebukuro K, Miyata A, Cho SJ, Nomura Y, Chang SM, Yamauchi Y, Hasebe Y, Uchiyama S, Karube I (1996) J Biotechnol 48:73
74. Hollis RP, Killham K, Glover LA (2000) Appl Environ Microbiol 66:1676
75. Billinton N, Baker MG, Michel CE, Knight AW, Heyer WD, Goddard NJ, Fielden PR, Walmsley RM (1998) Biosens Bioelectron 13:831
76. Rogers KR (2006) Anal Chim Acta 568:222
77. Sakaguchi T, Amari S, Nagashio N, Murakami Y, Yokoyama K, Tamiya E (1998) Biotechnol Lett 20:851

Adv Biochem Engin/Biotechnol (2010) 117: 21–55
DOI: 10.1007/10_2009_21

Published online: 21 January 2010

Mammalian Cell-Based Sensor System

Pratik Banerjee, Briana Franz, and Arun K. Bhunia

Abstract Use of living cells or cellular components in biosensors is receiving increased attention and opens a whole new area of functional diagnostics. The term "mammalian cell-based biosensor" is designated to biosensors utilizing mammalian cells as the biorecognition element. Cell-based assays, such as high-throughput screening (HTS) or cytotoxicity testing, have already emerged as dependable and promising approaches to measure the functionality or toxicity of a compound (in case of HTS); or to probe the presence of pathogenic or toxigenic entities in clinical, environmental, or food samples. External stimuli or changes in cellular microenvironment sometimes perturb the "normal" physiological activities of mammalian cells, thus allowing CBBs to screen, monitor, and measure the analyte-induced changes. The advantage of CBBs is that they can report the presence or absence of active components, such as live pathogens or active toxins. In some cases, mammalian cells or plasma membranes are used as electrical capacitors and cell–cell and cell–substrate contact is measured via conductivity or electrical impedance. In addition, cytopathogenicity or cytotoxicity induced by pathogens or toxins resulting in apoptosis or necrosis could be measured via optical devices using fluorescence or luminescence. This chapter focuses mainly on the type and applications of different mammalian cell-based sensor systems.

Keywords Biosensor • Pathogens • Toxins • Cytotoxicity • Rapid detection

---

P. Banerjee
Laboratory of Food Microbiology & Immunochemistry, Department of Food & Animal Sciences, Alabama A&M University, Normal, AL 35762, USA
Molecular Food Microbiology Laboratory, Center for Food Safety Engineering, Purdue University, West Lafayette, IN 47907, USA

B. Franz and A.K. Bhunia (✉)
Molecular Food Microbiology Laboratory, Center for Food Safety Engineering, Purdue University, West Lafayette, IN 47907, USA
e-mail: bhunia@purdue.edu

Contents

1 Introduction

The cell-based biosensor (CBB) systems that incorporate whole cells or cellular components have an obvious advantage of responding in a manner that can offer insight into the physiological (at organelle, cellular, or tissue levels) effect of an analyte [1]. Cell-based assays (CBA) are emerging as dependable and promising approaches to probe the presence of pathogens in clinical, environmental, or food samples [2–6]. Living cells are extremely sensitive to modulations or disturbances in "normal" physiological microenvironment. Therefore, CBBs are employed to screen and monitor "external" or environmental agents capable of causing perturbations of living cells [5, 6]. Some of the CBAs utilize the metabolic responses of cells (like cyanobacteria) to detect biological or chemical entities, like oxygen and herbicides in water [7]. In another type, mammalian cells or plasma membranes are used as electrical capacitors. The mechanical contact between cell–cell and cell–substrate is measured via conductivity or electrical impedance [8, 9]. Also, mammalian cells are used to measure biochemical and metabolic end-products delivered from cultured cells to the medium [6]. The CBAs can also measure the direct electrical response of electrogenic cells such as neural cells, heart muscle cells, pancreatic beta cells, or a neural cell network [6].

Pathogenic microorganisms must interact at a molecular proximity with cellular components of the host cells in order to elicit the pathogenic response. The mammalian plasma membrane is the first cellular interface for such an encounter with a microbe or microbial products like toxins. Most often, as a result of this encounter, the molecular and structural integrity of the plasma membrane is altered [10, 11].

Plasma membranes of mammalian cells may release membrane-bound markers when they are damaged by the action of a pathogen or toxin [12–14]. This interaction of host cell membrane and pathogen is archetypal [15]. The damage caused by a pathogen on plasma membrane bears a signature of the pathogen itself [15, 16].

In recent years, applications of functional detection of toxic and hazardous agents using CBBs integrating CBAs are reported extensively [2, 3, 17–20] and offer significant promise. Cell-based systems have proved their potential and emerged as some of the most significant approaches in drug development, toxicology, and high-throughput screening (HTS) [21]. Development of detection systems based upon the physiological action of pathogens and toxins (pathogenicity) is receiving increased attention in clinical pathology, as well as in food diagnostics, environmental monitoring, and biosecurity applications. Therefore, a biosensor capable of rapid and functional screening of toxins and pathogens in food samples is in great demand to cater to the contemporary needs of food safety and biosecurity.

The changing scenario of threats and issues of biosecurity and screening of classified agents related to food, agricultural, or environmental safety necessitates not only a sensitive detection regimen, but also requires a rapid, broad spectrum screening tool that can be employed as the "first line of defense." The nature and capability of the detection systems must conform to challenges associated with emerging pathogens and for unknown or little known biohazards. The present methods of detection depend mostly on known chemical characteristics or molecular recognition of target organisms, toxins, or substances [5]. In this regard, conventional detection systems fail to detect or identify unknown or emerging analytes, since most of these methods are specifically designed to identify a particular organism or toxin. In this context the need for a "functional detection" system is of utmost importance [22]. A functional detection system is capable of reporting the presence of threat agents or substances in a physiologically relevant manner. A living system can provide some of the key functional information about an analyte [5, 23] which is listed below.

1.1 Toxicity of the Analyte

This can be assessed by using living systems as sensing elements, such as bacteria, higher eukaryotic or mammalian cells, or living animals. The physiological responses, such as membrane damage, apoptosis, and necrosis of the sensing elements as a result of exposure to test substances can be deduced to gain information about the nature of toxicity of an analyte.

1.2 Impact on Cellular Metabolism

This can be evaluated by the alteration of signal transduction, second messenger pathways, or enzyme pathways.

1.3 Receptor–Ligand Interaction

A test substance might act as agonist or antagonist to a particular cell-type or groups of cells. Therefore, a living cell containing such receptors could serve as a better sensing probe to evaluate the agonistic or antagonistic potential of test substances.

1.4 Changes in Gene Expression

The interaction with foreign substances may alter the gene expression profiles of sensing cells, tissues, or organs as a result of exposure. Functional genomics of cultured cells can discern and classify agents or analytes based on their capabilities to intervene gene expression of the sensing cells.

2 Animals as Sensor

Ideally, the best sensing system with the capability of functional detection of human or animal threat agents would obviously be the living animals themselves. Different animals have been used as "sentinels" for detection of toxic and harmful agents (Table 1). In their two comprehensive reviews of the scientific literature from 1966 to 2005, Rabinowitz et al. [24] revealed evidence that animals can potentially provide early warning of an acute bioterrorism attack as well as serve as markers

Table 1 A time-line of examples of animals as sentinels of environmental toxicants

Sentinel incidents			Date	References
Species	Toxicant	Country		
Canaries	Carbon monoxide	England (Wales)	1870s	[168, 169]
Cattle	Smog	England	1910s	[170, 171]
Cattle	Fluoride	England		[26]
Horses	Lead	USA		[26]
Cattle	TCE	Scotland		[172, 173]
Cats	Mercury	Japan	1950s	[174]
Birds	DDT	USA		[26]
Chickens	PCB	Japan	1960s	[175]
Sheep	OP agents	USA		[176]
Horses and other animals	Dioxin	USA	1970s	[177, 178]
Dairy cattle	PBBs	USA	1980s	[179, 180]
Sheep	Zinc	Peru		[26]
Alligators	DDT, dicofol	USA	1990s	[181]
Fish	*Pfiesteria* toxins	USA		[26]

TCE trichloroethylene, *PCBs* polychlorinated biphenyls, *OP* organophosphate, *PBBs* polybrominated biphenyls
Source: van der Schalie et al. [26]

for ongoing human environmental health hazard exposure risks [24, 25]. The sentinel animal species have been successfully used for environmental monitoring applications, hazard and risk assessments, and detection and screening of deleterious changes in animal and human ecosystems [26].

Even though the use of animals could provide useful information about physiological risks associated with a particular agent or groups of agents for a prolonged period of time, the limited portability and robustness for practical large scale application for in-field or military applications preclude the use of large living animals for biosensing. The application of living cells to monitor and screen pathogens or toxic substances offers many advantages over using a whole animal for functional biosensing. Li et al. [27] listed several advantages of using cultured cells over using whole animals, such as, when cells are used for evaluation of different analytes. The key factors of cellular function affected by those analytes can be singled-out without interference from more complex, whole-organism or whole-organ responses. In addition, the cell types distributed or grown in a thin layer provide better optical observation capabilities using microscopes or other optical devices. The time necessary to raise an animal can be avoided and, most importantly, a wide range of medium formulations and cell lines originating from most tissue types are commercially available [27]. In addition to these factors, some other advantages of using cells can be attributed to their abilities of responding to external stimuli in a way that is physiologically relevant. Moreover, culturing and maintenance of cells are less expensive compared to animal maintenance. Also, CBAs provide much broader and more complex functional information, such as global information about protein synthesis and apoptotic or necrotic cell death when compared to nucleic acid and immunochemical methods. Information obtained by a CBA can provide insight into the mechanism of toxicant or pathogenic action, which in turn facilitates not only detection but also agent classification. Considering all these factors, "cell-based biosensors" (CBB) incorporating living cells, tissues, or cellular components have become very popular and useful in functional biosensing [2, 28]. As such, cell-based detection or screening of hazardous agents is an attractive approach, since it avoids the impracticalities associated with the use of whole animals for function-based detections or limitations of conventional biochemical or molecular detection strategies, such as nucleic acid or antibody-based methods (Fig. 1).

3 Cell-Based Biosensor

When used in detection systems, living cells or cellular components demonstrate unique capability of functional evaluation of the analyte and, therefore, physiological effect of the analyte can be deduced from the response [5]. Biosensors which incorporate CBAs have become a reliable and promising approach to investigate the presence of hazardous agents (such as pathogens or toxicants) in clinical, environmental, or food samples [5, 6]. The CBB systems can screen and detect

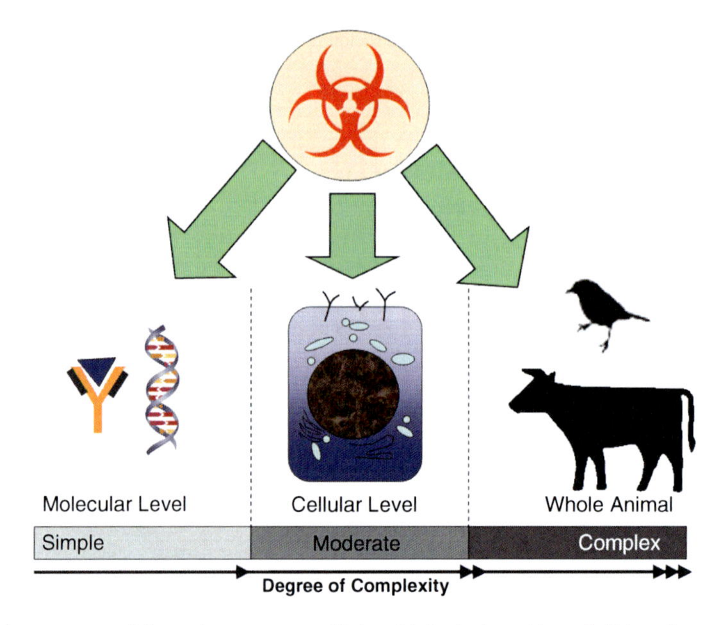

Fig. 1 Assessment of hazardous agents utilizing biological entities. Cell-based assays offer function-based diagnostics with optimal complexity

anomalies, disturbances, or changes in "normal" physiological activities of mammalian cells initiated by an "external" or environmental agent [5, 6].

In some of the CBB applications, mammalian cells or plasma membranes derived from cells have been used as electrical capacitors or the mechanical contact between cell–cell and cell–substrates has been measured by evaluating the conductivity or electrical impedance [8, 9, 29]. Biological products in water, such as herbicides or oxygen (as a metabolic end product) were detected using the physiologic responses of cells (e.g., cyanobacteria) [7]. Biochemical and metabolic end-products released by cultured cells into the medium in response to analytes were measured by CBBs connected to pH sensors or potentiometers [6]. Electrogenic cells, such as neural cells, heart muscle cells, pancreatic beta cells, or neuronal cell networks are used extensively in CBB applications to measure the direct electrical response of these cells when exposed to different agents [5, 6].

To overcome issues with stability and robustness of using living cells in sensor applications, cellular components such as lipid membranes are being used in CBB. Cellular components derived from whole cells are more stable than the whole cell itself and they eliminate the need for maintaining expensive cell culture setup [30–33]. Components of cell membranes such as receptors, e.g., G-protein-coupled receptors (GPCR), ion channel receptors, or cell surface proteins such as heat-shock proteins are immobilized onto functionalized sensor surfaces for cell-based detection assays [4, 34, 35].

3.1 Bacterial Cells vs Mammalian Cells

The term "cell-based biosensor" has been equally implicated in the description of detection systems employing bacterial or prokaryotic cells as well as higher eukaryotic or mammalian cells [1, 6, 36]. There have been several reports where whole cell microorganisms, such as bacteria and algae, were used as the recognizing elements in detection of environmental hazards such as heavy metal ions [37, 38]. Using photosynthetic green algae and cyanobacteria, Sanders and co-workers reported sensitive detection of the nerve agents tabun (GA), tributylamine, and dibutyl sulfide [39]. A microbial amperometric biosensor based on yeast cells was used for detection of lactate [40]. In another type of bacterial CBB, Immonen and Karp [41] employed two genetically engineered *Lactobacillus lactis* ssp. *cremoris* strains for detection of nisin in food samples. A rapid and selective method for the determination of free short-chain fatty acids in milk was developed by Schmidt and co-workers using a microbial sensor based on thick film oxygen electrode technology [42]. Several groups have used bacterial bioluminescence-based optical biosensors or electrochemical sensors containing recombinant or wild type microbial cells in rapid environmental monitoring systems for onsite biochemical oxygen demand or toxicant evaluation [20, 43–47].

The advantages of using single-cell microorganism-based sensors include rapid growth and ease of culturing. Culturing of bacterial cells is usually less expensive than that of mammalian cell culture, and bacteria are more robust and have better viability than mammalian cells. Despite these advantages, the mammalian cells can report information such as bioavailability [48], cellular metabolism, and physiologic responses relevant to human and animals [9, 17, 49–52] or cytotoxic responses [3, 12, 17, 19, 53, 54]. This type of information is important for functional biosensing. The ability of mammalian cells to simulate the organism's physiologic function and interact with toxicants has become an essential tool for screening, sensing, and evaluating environment, food, or clinical hazards [3, 5].

4 Classification of Mammalian CBBs

A variety of mammalian CBBs exist which differ widely not only in types of cells utilized, but also in detection applicability and methodology used. Therefore, the employment of mammalian CBBs is applicable across many fields from medical diagnostics and drug discovery to pathogen and toxin testing. The great diversity in this field coupled with innovative advances in the engineering of detection platforms allows for a great heterogeneity in the composition and use of any individual mammalian CBB. CBBs are broadly classified based on the function and mechanism of action of the biosensing (Table 2).

Table 2 Classification of cell-based biosensors

Function/mechanism at cellular levels	Cell types	Primary signals derived from cells	Device/methods used for secondary transduction of signals
Excitable/electrogenic	Neuron, cardiac cells, neuronal network	pH, flow of ions	LAPS, microelectrodes
Electrical responses	Epithelial cells, cardiomyocytes, neuron	Electric current, flow of ions	Impedance, IDES
Cellular receptors	Epithelial cells, hepatocytes, stem cells, mononuclear cells, T- or B-cells	pH, alteration of molecules within cells	Cell-signaling molecules, LAPS, optical methods
Cellular metabolism	Epithelial cells, hepatocytes, stem cells	pH, ion channel, molecular flux	pH-sensitive ISFETs, LAPS, ion-sensors
Cytotoxicity	Epithelial cells, endothelial cells, macrophages, myeloma, mononuclear cells, T- or B-cells	Changes in membrane integrity, cellular morphology	Optical methods, potentiometric methods
Genomic responses	Epithelial cells, endothelial cells, macrophages, myeloma, T- or B-cells	Changes in gene expression	Reporter gene assay, optical methods, cytometry

LAPS light-addressable potentiometric sensor, *IDES* interdigitated electrode structures, *ISFET* ion-sensitive field effect transistor

4.1 CBBs Utilizing Electrical Measurements

4.1.1 CBBs Utilizing Excitable or Electrogenic Cells

Bioelectric and chemical responses of excitable cells like cardiac myocytes [48], neurons [5], glial cells, cultured neuronal networks [5], brain tissue slices, or vertebrate retina [55, 56] can be utilized to deduce functional information of an analyte [1]. In this class of CBBs, stimuli (for example, chemicals, such as toxins, drugs, or physical stimulant, such as light or electronic signal) are added into a detection chamber containing cells cultured on a chip. As a result of stimulus, the cells produce action potential resulting from physico-chemical changes, such as pH of culture medium, which is detected by an electrical device through a thin layer of electrolyte [57, 58]. By evaluating the changes in electrophysiological activities of the cardiac myocyte cells from rats, Liu et al. [59] demonstrated sensitive detection of heavy metals, such as mercury, lead, cadmium, etc., using a light-addressable potentiometric sensor (LAPS) [48].

The use of microelectrodes for extracellular recording from electrogenic cells has become a popular technique for different applications ranging from toxicant evaluation to drug screening [60, 61]. To conduct extracellular recordings, commercial systems are available from different manufacturers, such as Plexon Inc. (Dallas, TX), Multichannel Systems GmbH (Reutilingen, Germany), and Alpha MED Sciences, Co. (Tokyo, Japan) [60, 62]. Researchers at the US Naval Research Laboratory in collaboration with University of North Texas reported the development of microelectrode array using cultured neuron for quantitative measure of synchronization in neuronal impulses [63–66]. The typical characters of neuronal spikes under exposure to different chemicals were correlated with type of exposure leading to a specific pattern for a specific group of agents.

4.1.2 CBBs Utilizing Changes in Impedance

Living cells grown on a surface (adherent cell types) function similar to a simple circuit, since they can be considered an electrical system containing conductive fluid packaged within a membrane surrounded by another conductive fluid. The conductive fluids make up the resistance elements of the circuit, while the membrane acts as a capacitor. Impedance methods have been used to monitor tissue cultures online and in real time [67]. Impedance-based CBBs utilize the ability of adherent cells to change electrical impedance due to the dielectric properties of membranes of biological materials including cells. Impedance-based measurements rely on the phenomenon that intact living cells are excellent electrical insulators at low signal frequencies. As cells grow or migrate to increase surface area of coverage over an electrode surface, the effective electrode impedance rises. Adherent cells grown on planar electrodes can be used as a CBB for impedance measurements to indicate cell adhesion, spreading, and motility, since it has been found that the changes in cellular motility and spreading can be associated with exposure to different analytes [9, 29].

In a particular type of impedance-based CBB, cells derived from a male monkey's kidney were adherently grown on interdigitated electrode structures to monitor impedance changes associated with cell growth [68, 69]. The cellular behaviors of nonexcitable cells like endothelial cells [70], fibroblasts [9], or macrophages [71] were monitored using impedance measurements. In another type of impedance-based CBB, Kovacs and colleagues at Stanford University and at the US Naval Research Laboratory achieved functional agent sensing by monitoring changes in membrane impedance [72, 73]. Prolonged cellular growth period and morphological changes have been implicated as disadvantages of impedance-based biosensing.

4.2 CBBs Utilizing Biological and Physiological Measurements

4.2.1 CBBs with Cellular Receptors

One of the salient features of CBBs is that the cellular events initiated by analytes or agents can be interpreted and explicated both for detection as well as to deduce functional information about the mechanism of action of the analyte. A typical cell–analyte event is initiated by the intimate interaction of cellular receptors (most often referred to as "membrane receptors") with the foreign agents (analyte or ligand). Receptors may serve as the "entry port" for the foreign materials into the cell. For example, mammalian membrane proteins act as receptors for bacterial surface proteins and aid in the internalization of the bacteria [74] or may induce downstream intracellular signaling pathways upon binding to membrane receptors (for example, GPCRs) [75]. In-depth information about biosensing strategies using natural cellular receptors are reported elsewhere [76, 77]. Since most CBBs directly or indirectly exploit cellular receptors as the first biorecognition element, the receptors warrant significant attention in the field of development of cell-based biosensing strategies. A "rational design" of CBBs can become much easier when enough information about types of receptors expressed by a particular cell-type is available, and that information can effectively be utilized for pathogen or hazardous agent detection.

4.2.2 Cellular Receptors Used for Biosensing and Their Types

Cellular receptors are proteins (glycoproteins or lipoproteins) typically associated with cell membrane (or within the cytoplasm or cell nucleus) that bind to molecules or chemical entities (ligands) such as proteins and hormones, and upon binding the structural conformation of the receptor changes, resulting in the initiation of the cellular response. As such, ligand-induced changes in the behavior of receptor proteins results in physiological changes in the cell that represent the biological effect of the ligand. In a receptor-based biosensor, the typical cellular responses initiated by the ligand (in this case the analyte) can be interpreted to identify the analyte (detection), since receptor–ligand interactions are often specific or at least provide information about structural similarities about a group of analytes. For example, estrogen receptors on a cell bind to estrogen-like foreign substances called xenoestrogen (having structural similarity to estrogen and considered as environmental pollutants) and elicit estrogenic responses in estrogen responsive cell types [78].

Based on structural and functional characteristics, membrane receptors were classified into four major categories, namely, ion channel receptors, G-protein-linked receptors, receptors with a single transmembrane domain, and enzyme-linked receptors [76, 79], and are summarized in Table 3.

Ion Channel Receptors

Ion channel receptors, ionotropic receptors, or ligand-gated ion channels (LGIC) are responsible for the physiological activities which coordinate information flow in the brain and control behavioral activities. Any disturbance in the control of this balance leads to deleterious outcome, resulting in abnormal activity (e.g., epilepsy), behavioral disorders (e.g., anxiety), or even neuronal cell death (excitotoxicity) [80]. The LGICs activated by extracellular ligands may be divided into four superfamilies: (1) *Cys-loop superfamily*, (2) *glutamate receptors* (NMDA, *N*-methyl-D-aspartate; AMPA, α-amino-3-hydroxy-5-methyl-4-isoxazolepropionic acid; and

Table 3 Classification of membrane receptors

Characteristics	Receptor types			
	Ion channel receptors	G-protein-linked receptors	Receptors with a single transmembrane domain	Enzyme-linked receptors
Endogenous ligands	Neurotransmitters	Neurotransmitters, hormones, autoacoids, chemotactic factors	Growth factor hormones, cytokines	Atrial natriuritic peptide ligands, growth factors
Structure	Several proteins with a pore	1–2 proteins	1–2 proteins with catalytic domain	Individual protein linked with enzyme
Transmembrane segments	Four	Seven	One	Single-pass transmembrane proteins
Function	Regulation of ion transport	Activation of G-proteins, regulation of cellular functions and expression of proteins	Catalytic	Suppress proliferation, stimulate synthesis of extracellular matrix, stimulate bone formation, attract cells by chemotaxis
Cellular responses	Depolarization/ hyperpolarization	Depolarization/ hyperpolarization	Regulation of cellular functions, proliferation, and differentiation	Regulation of cyclase, production of cyclic GMP, cell signaling and regulation of cell cycle

Source: Subrahmanyam et al. [76]

kainite), (3) *TRP (transient receptor potential) channels*, and (4) *ATP-gated channels*. In mammals, the Cys-loop superfamily comprises cationic receptors (nicotinic, 5-HT, and zinc activated) and anionic receptors [γ-aminobutyric acid (GABA) and glycine receptors]. The members of this superfamily exhibit important physiological functions and mutations may lead to a range of pathological states. Ligands specific for this class of receptors include endogenous chemicals such as neurotransmitters. Typical examples of these include GABA, glycine, serotonin, and ATP.

G-Protein-Coupled Receptors

A vast array of cellular signaling pathways are mediated through GPCRs, interacting with numerous signaling molecules of diverse structure and function, including hormones, neurotransmitters, and local mediators. GPCRs consist of a single polypeptide chain spanning the lipid bilayer which includes several transmembrane segments. Endogenous ligands belonging to this class that are important target analytes for sensor technology include all neurotransmitters, most hormones and autocoids, chemotactic factors, and exogenous stimulants such as odorants [76]. For example, Whitaker and Walt [81] recently developed a fiber-optic single-cell array-based biosensing method using Chinese hamster ovary (CHO) cell line ectopically expressing different human GPCRs to detect and analyze real-time Ca^{2+} responses resulting from exposure to various agonists.

Receptors with Single Transmembrane Segments

Some of the crucial cell-signaling pathways are regulated by a group of receptors with single transmembrane segments. The prominent members of this group include growth factors, such as epidermal growth factors, platelet-derived growth factors, fibroblast growth factors, and nerve growth factors. Structurally these receptors are segmented into three domains, an extracellular domain responsible for ligand binding, a single transmembrane segment responsible for signal transmittal and conformational events, and cytoplasmic domains eliciting cellular responses by activating signal transduction pathways. Proteins such as phosphorylase C, GTPase-activating protein, and phosphatidylinositol 3-kinase are some classical examples of this receptor type.

Enzyme-Linked Receptors

Enzyme-linked receptors are probably the most studied group of receptors pertaining to pharmacological and clinical applications. These receptors are also transmembrane proteins having ligand-binding domains on the outer surface of the plasma membrane, while the signaling elicitor component may or may not be physically associated with the transmembrane domain. Broadly, these receptors are grouped into five major categories (adopted from Subrahmanyam et al. [76]) as follows. (1) *Receptor guanylyl cyclase* – ligand specificity is peptide hormones secreted by heart muscle cells. (2) *Receptor tyrosine kinase* (RTK) – this is a high affinity cell surface receptor containing a large transmembrane domain and a glycosylated extracellular domain. There are approximately 20 RTK families based on their affinity to different ligands [82], such as epidermal growth factor family, insulin receptor family, fibroblast growth factor family, vascular endothelial growth factor family, and others. (3) *Cytokine receptor superfamily* – this group includes growth hormone prolactin and antigen-specific receptors on T- and B-lymphocytes that regulate proliferation and differentiation in the hematopoietic system [83]. (4) *Tyrosine phosphatases* – these interact with phosphotyrosines on a particular type of protein and play significant roles in cell signaling. (5) *Serine/threonine protein kinases* – these perform various functions such as suppression of proliferation and stimulation of synthesis of extracellular matrix (ECM). Phosphacan, a chondroitin sulfate proteoglycan of nervous tissue, is a typical example of ligand for this class of receptors [84].

4.2.3 Receptors for Host–Pathogen Interaction

Multicellular organisms express specialized surface receptors called "cell adhesion receptors" so as to form tight associations with neighboring cells and with ECM to facilitate building cell layers, tissues, and organs. Cell adhesion receptors are constituted by different protein families, most important of these are the integrins, the cadherins, the immunoglobulin superfamily cell adhesion molecules (IgCAMs), the selectins, and the syndecans [85]. It is well established that, in addition to their structural role, adhesion receptors in most cases also play a crucial role in signal transduction from the exterior to the interior of the cell [86]. As a result of these multiple functions, the expression of adhesion receptors on the cell surface is critical and integrities of these proteins are conserved in multicellular organisms. However, due to the nature of the surface exposure, signaling capacity, and conservation, these receptor proteins have evolutionarily became ideal targets for pathogen interaction, communication, and invasion [85, 87].

Several ECM proteins and adhesion receptors are targeted by different pathogenic microorganisms (Table 4). Three major classes of these receptors – integrins, cadherins, and IgCAMs – are discussed below with regard to their interactions with pathogenic microorganisms.

Integrin Receptors and Interacting Pathogenic Bacteria

Integrins are cell surface receptors that are glycoprotein in nature and interact with ECM proteins or recognize membrane-bound counter receptors and elicit intracellular signal transduction. Some outer membrane proteins expressed by enteropathogens

Table 4 Pathogens targeting cell adhesion molecules

Species	ECM protein/receptor
Borrelia burgdorferi	FN/β1 integrins
Mycobacterium leprae	FN, LN/β1 and β4 integrins
Mycobacterium bovis BCG	FN/β1 integrins
Neisseria gonorrhoeae and *N. meningitidis*	FN, VN/β1 and β3 integrins
Porphyromonas gingivalis	β1 integrins
Shigella flexneri	β1 integrins
Staphylococcus aureus	FN, LN, Col/β1 integrins
Streptococcus pyogenes and *S. dysgalactiae*	FN/β1 integrins
Yersinia pseudotuberculosis and *Y. enterocolitica*	β1 integrins
	Immunoglobulin-related cell adhesion molecules (IgCAMs)
Haemophilus influenzae	CEACAMs
Moraxella catarrhalis	CEACAMs
Neisseria gonorrhoeae and *N. meningitidis*	CEACAMs
	Cadherins
Listeria monocytogenes	E-cadherin

CEACAMs carcinoembryonic antigen-related cell adhesion molecules, *Col* collagen, *FN* fibronectin, *LN* laminin, *VN* vitronectin
Source: Hauck et al. [85]

like *Yersinia* species, *Y. pseudotuberculosis* and *Y. enterocolitica*, and *Shigella* (IpaC protein) function as ligands for mammalian β1-integrin receptors. Gram-positive bacteria such as *Staphylococcus aureus* or *Streptococcus pyogenes* interact with integrins by recruiting ECM proteins. These ECM proteins are adhesins in nature and include fibronectin-binding proteins A and B (FnBP-A and -B) of *S. aureus* or Sfb1 (also called F1) of *S. pyogenes* [88, 89]. Enteropathogenic *Escherichia coli* (EPEC) interacts with β1-integrins through its surface intimin adhesin [90]. EPECs insert a protein into the host cell membrane called transepithelial intimin receptor (Tir) which serves as the counter receptor for the bacteria-associated intimin [91]. Evidence from in vitro experiments using cultured cells indicates that the region of Tir that is recognized by intimin displays homologies to β1-integrin receptors, leading to the belief that intimin might also bind to certain cellular integrins [92].

Immunoglobulin Superfamily Cell Adhesion Molecules and Associated Pathogenic Bacteria

Cell adhesion molecules of the immunoglobulin superfamily are characterized by the presence of at least one immunoglobulin-like domain in their extracellular part. The immunoglobulin domain functions as a site for binding and recognition. IgCAMs appear as integral membrane proteins or coupled to the membrane via a glycosyl-phosphatidylinositol anchor. The prominent pathogen exploiting IgCAMs is found in the genus *Neisseria*, where pathogenic *N. gonorrhoeae* and *N. meningitidis* both express adhesins, the colony opacity-associated (Opa) proteins that bind to members of the carcinoembryonic antigen-related cell adhesion molecule (CEA-CAM) family, a subgroup of IgCAMs that elicits the pathogenesis of this microorganism [85, 87, 93].

Cadherins and Interacting Pathogens

Cadherins are transmembrane glycoproteins that mediate tight homotypic cell–cell association [94]. Cadherins possess five cadherin-motif subdomains in their extracellular segment which allow them to dimerize with neighboring cadherin molecules in the presence of Ca^{2+} [95]. *Listeria* is considered to be the most prominent example of pathogenic bacteria engaging cadherins. *Listeria monocytogenes*, a Gram-positive, facultative intracellular pathogen, expresses a family of adhesins termed internalins that mediate the invasion of this pathogen into different cell types [96]. Cossart's group at the Pasteur Institute, France has led the way in understanding the *Listeria* surface proteins, such as internalins and their interactions with host cell receptors. Two types of internalins are proposed to be the key virulence factors playing important roles in the host cell adhesion–invasion mechanism for *Listeria*: Internalin B (InlB) aids in the invasion into a number of different cell types and seems to associate with different receptors such as Met receptor tyrosine kinase [97, 98], while internalin A (InlA) allows the bacteria to penetrate efficiently human epithelial cells [99]. This property is based on InlA binding to human E-cadherin [100]. Interestingly, like the homophilic interaction between cadherins, the binding of InlA to E-cadherin is Ca^{2+} dependent. As cadherins are linked to the actin cytoskeleton, it is not surprising that InlA-mediated invasion can be blocked by agents that disrupt the integrity of the actin cytoskeleton such as cytochalasins [101, 102].

Other Mammalian Surface Proteins

The intimate physical interaction between host–pathogen is mediated by several other surface expressed molecules on mammalian cell membrane. Prominent examples of this are the group of proteins called heat-shock protein (Hsp). Our group has discovered that under thermal stress conditions *Listeria monocytogenes* can over-express a 104 kDa surface alcohol acetaldehyde dehydrogenase designated Listeria adhesion protein (LAP), which acts as a ligand to a ubiquitous mammalian heat-shock protein (Hsp60) and aids in bacterial adhesion and translocation in intestinal epithelial cells [103–105]. Interestingly, physiological stressors and nutritional status can up- or downregulate the expression of LAP, and thereby its interaction with its receptor Hsp60 [106, 107]. Viruses also utilize receptors to gain entry into host cells. These may include one or more attachment receptors and at least one entry receptor [108]. Interaction with attachment receptors increases infectivity but may be an optional requirement [109, 110]. Examples of attachment receptors include cell-bound heparan sulfate molecules for HIV [110] and herpes simplex virus [111]. The entry receptors for HIV are CD4 and members of the extended chemokine receptor (CCR) family (also known as co-receptors) [112]. Some viruses use multiple alternative receptors to enter a host cells. For example, HIV uses CCR5, or CXCR4, CCR2b, CCR3, CCR8, and CCR9 to enter host cells. The repertoire of host cell surface receptors available to a given virus comprises the receptor array. Specific interaction of virus with the host cell depends on the availability of a minimal set of receptors from a series of available receptors in that host cell surface [108].

4.2.4 Application of Receptors in Functional Biosensing

The application of receptors for cell-based sensing is a technique which has been used for a long period of time. CBAs for HTS are a widely used technique in drug discovery and in clinical/pharmacological applications [21]. In recent years, interfacing of electronics and optical technologies has aided in technological advancement and instrumentation for CBAs. Some examples of pioneering technologies using receptor-based HTS, utilizing GPCRs [113], and ion channels [114, 115] are FLIPR (Molecular Devices) and Photina (PerkinElmer). Both are automated systems for measuring calcium concentrations in cellular compartments. The former uses fluorescence imaging while the latter utilizes luminescence assays. Other optical tools such as confocal imaging and laser scanning microscopy and cytometry enable acquisition of real-time and time-lapse cellular and sub-cellular images. Laser scanning imaging utilizing fluorescence imaging techniques is now routinely used for HTS and to evaluate receptor–ligand interactions [116].

Several automated systems are now available to monitor receptor–ligand interaction. FMAT 8100 HTS (Applied Biosystems) is a system which employs a macroconfocal scanning platform to visualize and quantify protein–peptide or protein–protein interactions in real-time [117]. ArrayScan (Cellomics) is used to perform high-content cell-based imaging assays to monitor cellular localization of molecules, cell–cell interactions, and cytotoxicity or receptor (GPCR) internalization [118–121].

4.3 CBBs Based on Metabolic Measurements

The metabolic activities of a cell are a function of its interactions with surrounding environment or agents. The changes to the extracellular environment due to cell metabolism (as a result of exposure to physical or chemical factors) can be measured. Manifestation of cellular metabolism is associated with changes in the rate of acidification of the media and changes in medium composition or characteristics, such as changes in extracellular pH or concentrations of different ionic species initiated by secreted cellular metabolites, such as glucose, lactate or ammonia [122]. Other common methods for metabolic sensing using living cells include oxygen and/or carbon-dioxide consumption or production and microcalorimetry [23]. Well-established biochemical sensors such as glass electrodes, pH-sensitive ISFETs (ion-sensitive field effect transistors), or LAPS can be utilized to monitor pH-changes, Clark electrodes for oxygen, and amperometric (enzyme sensing) devices for glucose and lactose monitoring [5, 6, 67]. Ion-selective ISFET can be used to monitor the concentrations of the ionic species. Such a potentiometric sensor was used to detect the effect of histamine as a model toxin on human umbilical vein endothelial cells [123]. With the advent of metabolomics, CBAs are receiving increased attention to screen various toxicants. For example, hepatocytes express high levels of drug-metabolizing enzymes (both phases I and II) and have been widely used in CBA platforms to evaluate the hepatotoxic potentials of various drug molecules or environmental toxicants [124, 125].

In another type of cell-based sensing platform, stem cell-based systems are used to evaluate agents which affect cell metabolism. Stem cell-based systems offer a promising and innovative alternative for obtaining large numbers of cells for early efficacy and higher toxicity screening, allowing improved screening of drug or toxicant molecules. The applications of adult and embryonic stem cell-based screening are popular choices in several areas of HTS including cardiotoxicity [126], hepatotoxicity [127], genotoxicity/epigenetic [128] and reproductive toxicology [124]. In another study, embryonic stem cells derived from a mouse were induced to differentiate in vitro to cardiomyocytes or neurons and were cultured on the surface of LAPS in order to monitor changes in extracellular ion concentrations [48]. The metabolic measurements offer functional information relevant to the analyte. At the same time, cross-talks between metabolic pathways may preclude specific detection of a particular analyte, since different toxicants or pathogens evoke different cellular responses, such as signaling pathways or targeting of different organelles.

4.4 CBBs Based on Optical Measurements

Optical assays utilizing the properties of various probing molecules such as color, fluorescence, or luminescence are utilized to acquire real-time and in-situ information about cell–cell or cell–analyte interaction. With the advent of novel organelle specific fluorophores it is possible to visualize a particular cellular component such as nucleic acids, mitochondria, or lipid membranes [129]. CBAs based upon fluorescent or luminescent properties for HTS application are probably the most powerful tools available to cell-biologists today [116, 120, 121, 130]. There have been several other optical methods employed to detect biological agents by monitoring cell death events. Cytotoxic agents which cause cell damage can be screened by using suitable markers for cytotoxicity. The hallmarks of cell death are membrane damage or nucleic acid degradation which eventually leads to either apoptotic or necrotic cell death [131, 132]. The quantification of cell death can be done by using appropriate enzyme substrate or fluorescent dyes [3, 12]. Several bacteria or mammalian cells have been genetically engineered to emit light (luminescence) when exposed to toxic compounds or pathogens [133–136]. In another type of optical assay cell viability was measured by monitoring respiration by optical oxygen sensing using a fiber-optic phosphorescent phase detector [137].

4.5 Genomic Biosensors

Genomic biosensors perform functional analysis of agents or drugs based on gene expression profiles in target cells [138]. A detailed review on genomic biosensors can be found elsewhere [139]. A genome based biosensor, such as GenomeScreen (developed by Aurora Biosciences Corp.), scans the genome broadly and identifies relevant genes in live cells. A library of cell clones is generated with a randomly integrated promoterless fluorescent reporter gene (such as β-lactamase) throughout the genome. A fluorescence activated cell sorting step is performed on this "Gene-Tag" library to isolate clones (of cells) based on the expression levels of individual tagged genes. This method is very specific and can provide information at the gene level; however, limited choice of appropriate cell lines with intact signal transduction pathways precludes wide use of this type of CBB.

4.6 CBBs Based on Cytopathogenicity

CBAs incorporating cytotoxicity measurements are gaining increasing importance in sensor design and detection strategies. The iniquitous distinction of "pathogens" is designated to the microorganisms which have the capability of inducing diseased conditions in other organisms such as humans, animals, and plants. The hallmark of the pathogenesis by a pathogen is the disruption of a normal physiological condition of the host. Such pathogenic organisms can be detected by damage inflicted upon the host system. Several pathogens are known to cause damage of varying severity to human and animal cells or tissues. A comprehensive list of such pathogens and their target cells is illustrated in Table 5.

4.6.1 Cytopathogenicity Assay for Pathogenic Bacteria or Toxins

Impairment of a host cell by pathogenic microbes or toxins often bears a "fingerprint" of that particular pathogen or toxin. These types of assays not only testify the presence of particular pathogens or toxin but also provide information about the functionality of the pathogen or toxin. Cytopathogenicity assays furnish critical information about the biological and physiological interactions occurring between host and pathogen (Fig. 2). Therefore, these cell-based cytotoxicity assays can distinguish a viable pathogen from a nonviable one. This information is often

Table 5 Human or animal cell lines used for cytopathogenicity assay for foodborne bacteria

Bacteria	Cell line	Cell type	Source	Cytopathic effects
Salmonella	CHO	Epithelial	Chinese hamster ovary	Elongation, detachment
	Vero	Fibroblast	Monkey kidney	Lysis, protein synthesis inhibition
	HEp-2	Epithelial	Human laryngeal	Invasion
	JY	B-cell	Human	Invasion
	H9	T-cell	Human	Invasion
	Henle-407	Epithelial	Human jejuna	Intracellular growth
	J774	Macrophage	Mouse	Intracellular growth
	HeLa	Epithelial	Human cervix	Toxicity, actin polymerization
	HT-29	Epithelial	Human colon	Apoptosis
E. coli	Vero	Fibroblast	Monkey kidney	Lysis, protein synthesis inhibition
	CHO	Epithelial	Chinese hamster ovary	Lysis, toxicity
	Henle-407	Epithelial	Human jejuna	Adhesion
	HEp-2	Epithelial	Human laryngeal	Adhesion, toxicity
	MAC-T	Epithelial	Bovine mammary gland	Invasion
	MDBK	Epithelial	Bovine kidney	Invasion
	HeLa	Epithelial	Human cervix	Toxicity, apoptosis
	T84	Epithelial	Human colon	Apoptosis
	Y1	Epithelial	Human adrenal gland	Rounding, detachment, cAMP
	W138	Fibroblast	Human lung	Toxicity
	J774	Macrophage	Mouse	Apoptosis
	HT-29	Epithelial	Human colon	Apoptosis
Shigella	HeLa	Epithelial	Human cervix	Cell death, protein synthesis inhibition
	Vero	Fibroblast	Monkey kidney	Lysis, protein synthesis inhibition
	3T3	Fibroblast	Mouse	Invasion, actin polymerization
	U937	Monocyte	Human	Apoptosis
	Mϕ	Macrophage	Mouse	Invasion, apoptosis
Campylobacter	HeLa	Epithelial	Human cervix	Distended cells (CDT effect)
	Vero	Fibroblast	Monkey kidney	Distended cells (CDT effect)
	AZ-521	Epithelial	Human stomach	Vacuolation
	Ped-2E9	B-cells	Mouse	Toxicity
Yersinia	J774	Macrophage	Mouse	Exocytosis, apoptosis
	HEp-2	Epithelial	Human laryngeal	Invasion, lysis
Vibrio	CHO	Epithelial	Chinese hamster ovary	Elongation, cAMP accumulation
Listeria	Caco-2	Epithelial	Human colon	Adhesion, invasion, apoptosis
	CHO	Epithelial	Chinese hamster ovary	Detachment, lysis
	Henle-407	Epithelial	Human jejuna	Intracellular growth, death
	Vero	Fibroblast	Monkey kidney	Toxicity

(*continued*)

Table 5 (continued)

Bacteria	Cell line	Cell type	Source	Cytopathic effects
	HEp-2	Epithelial	Human laryngeal	Invasion
	J774	Macrophage	Mouse	Intracellular growth
	RAW	Macrophage	Mouse	Intracellular growth
	HeLa	Epithelial	Human cervix	Toxicity
	HUVEC	Endothelial	Human umbilical vein endothelial cell	Intracellular growth
	Hep-G	Epithelial	Human liver	Intracellular growth, apoptosis
	3T3	Fibroblast	Mouse	Invasion, plaque formation, lysis
	L-M	Fibroblast	Mouse	Invasion, plaque formation, lysis
	NS1	Myeloma	Mouse	Lysis, toxicity
	Ped-2E9	B-cell	Mouse hybridoma	Lysis, apoptosis
	RI-37	B-cell	Human–mouse hybridoma	Lysis, apoptosis
	Ramos RA1	B-cell	Human	Lysis, apoptosis

Source: Bhunia et al. [4]

very important from an industrial (especially food manufacturing) point of view, as nonviable bacterial cells typically cannot cause disease and are excluded from the purview of safety regulations. In contrast, immunological and nucleic acid based methods for detecting pathogens and toxins primarily rely on chemical properties or molecular recognition to identify a particular agent. As a result, no functional (biological and physiological) information can be obtained from such assays.

4.6.2 Cytotoxicity Assay

The cytopathogenicity of a pathogen or toxin is revealed by the damage of the target host cells or tissues incurred by host–pathogen interaction. The common outcomes of a bacteria or toxin mediated injury to a eukaryotic cell are manifested by cell membrane damage or pore formation, cell death via apoptosis or necrosis, cell lysis, or detachment of cells from substrata (Fig. 2). The effect conveyed by pathogenic microorganisms or their toxins can be assayed using cytopathogenicity or cytotoxicity assays [4, 67]. A typical cytopathogenicity or cytotoxicity assay can be performed either directly by microscopic methods or by indirect determination of eukaryotic cell membrane damage or pore formation. The direct microscopic evidence, such as rounding of Vero cells, is routinely used to evaluate the presence of certain pathogens or toxins [140, 141]. Membrane damage can be measured by incorporation of dyes like trypan blue or propidium iodide [12, 53, 54, 142]. Diffused enzymes, such as alkaline phosphatase (ALP) or lactate dehydrogenase that are released from eukaryotic cells as a result of membrane damage, can also be assayed colorimetrically or spectrophotometrically to assess the extent of the cell-damage [12, 19]. However, there are limitations to assays probing for released

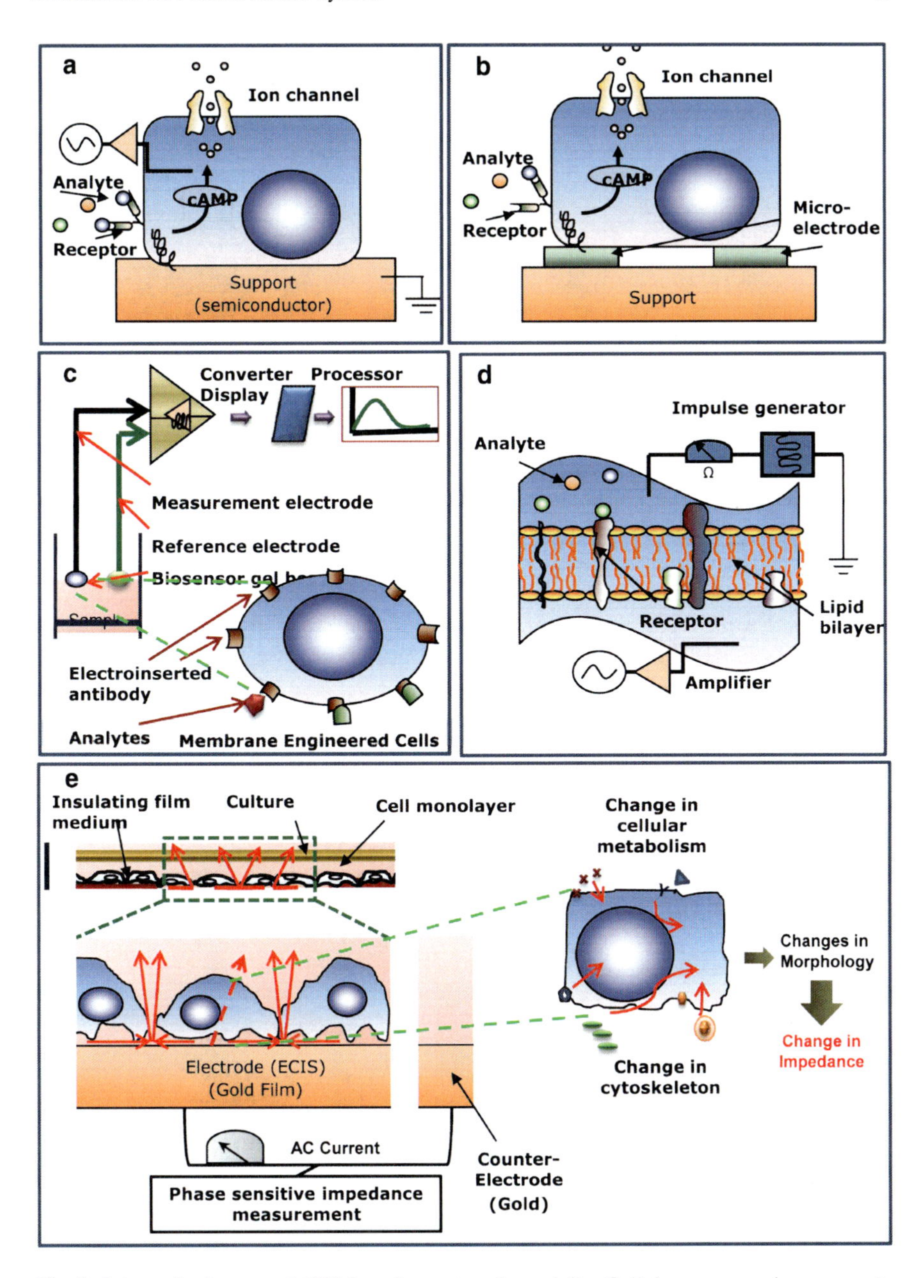

Fig. 2 Schematic diagram of CBB based on cytopathogenicity. Cellular processes in response to pathogens or toxins (external stimuli) can be exploited for evaluation of cytotoxicity. In a CBB system based on cytopathogenicity the cells release extracellular that is enzymes, such as alkaline phosphatase or lactate dehydrogenase or intracellular molecules, such as calcium ions. These marker molecules are detected as estimators of cytotoxicity

enzymes. These enzymes may not be present in all types of cells, and the effect of a certain membrane-degrading pathogen or toxin may not produce pores large enough to release more sizeable proteins. Other assays such as 3-[4,5-dimethyl thiazolyl-2]-2,5-diphenyltetrazolium bromide (MTT) or 4-[3-(4-iodophenyl)-2-(4-nitrophenyl)-2*H*-5-tetrazoliol]-1,3-benzenedisulfonate (WST-1) can report pathogen or toxin mediated cytotoxicity based upon parameters of proliferation, viability, or activation [143–145].

4.7 CBBs Using a 3D Cell Culture System

Mammalian cells cultured in three-dimensional (3D) configurations are considered a major breakthrough in biosensor design. Cells are grown in a biocompatible scaffold such as collagen, matrigel, hydrogel, or aliginate gel to provide a 3D architecture emulating tissues in the body [146]. Cells grown in 3D configuration more accurately represent gene expression, cell differentiation and other biological activities similar to in vivo models than 2D culture system [147–149]. The 3D cell culture system has been used to study host-cell–pathogen interaction for several microorganisms including *Salmonella* Typhimurium [150], uropathogenic *E. coli* [151], *Pseudomonas aeruginosa*, cytomegalovirus, and norovirus [152, 153]. Recently, collagen encapsulated hybridoma B-cells in 3D scaffold were successfully used in an array format in 96-well plate for detection of *Listeria monocytogenes* and *Bacillus cereus* toxins [3]. The 3D cell culture model shows much promise to be an integral part of the cell-based biosensor for HTS in pathogen testing, drug discovery and point-of-care applications [152, 154].

4.8 Current Status of Cell-Based Biosensing

The applications of biosensors that incorporate whole mammalian cells to detect foodborne pathogens are drawing increasing importance. This is because mammalian CBBs have the unique potential of distinguishing viable bacterial cells from those that are nonviable. As mentioned previously, this serves as a critical parameter in the food industry, since nonviable bacterial cells or inactive toxins do not raise any concern. The development of a CBB which could be used in the food industry is still undergoing rigorous laboratory validations. Some of the key developments in this area are listed below.

4.8.1 Electrical Cell–Substrate Impedance Sensing

"Electrical cell–substrate impedance sensing" (ECIS) was developed by Giaever and Keese [9] for the detection of changes in cellular morphology (Fig. 3e). In ECIS,

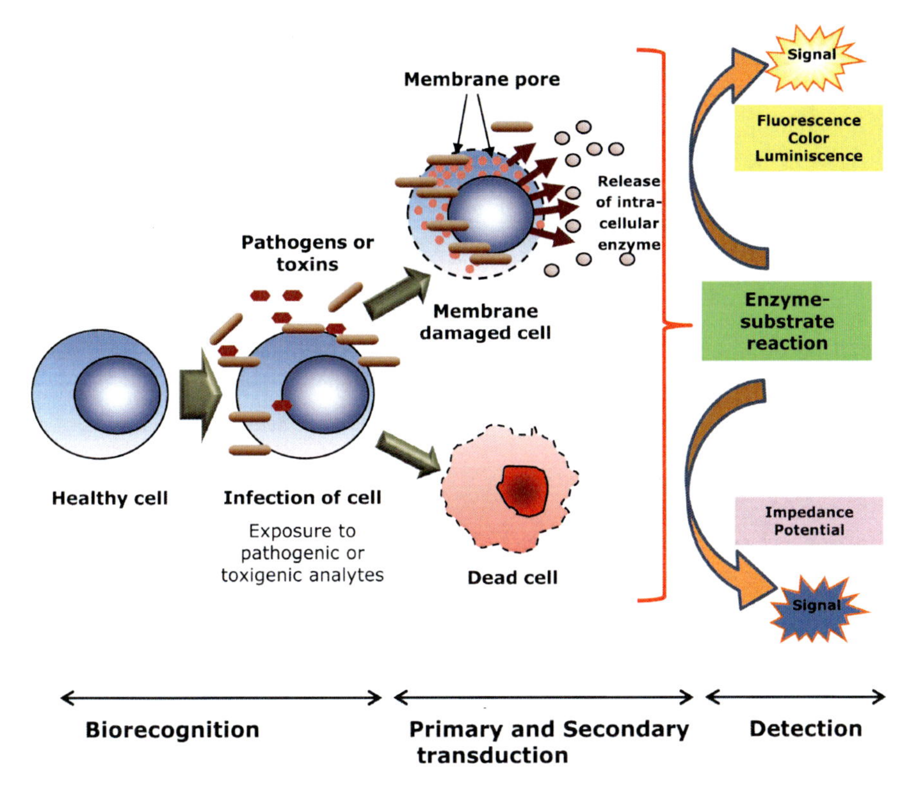

Fig. 3 Different strategies of construction of cell-based biosensors exploiting changes in electrical properties. (**a**) CBB that measures changes in electrical potential. An analyte interacts with a receptor located in the cell membrane. After interaction, an ion influx is caused by a signal cascade involving cyclic-adenosine monophosphate (*cAMP*). Changes in the electric potential of the cell are measured by a silicon-based semiconductor, which is located inside the support. The cell may be stimulated by a microelectrode. (**b**) Same cell as shown in (**a**) but the cell is immobilized on microelectrodes. Semiconductor structures may also be integrated into the support. (**c**) An immobilized CBB based on the bioelectric recognition assay (BERA). The measuring electrode is inserted in the immobilized cell-gel bead, while the reference electrode is inserted in a cell-free gel bead. The cell-gel beads contain mammalian cells those were engineered to carry analyte-specific antibody in the membrane at high density on the cell surface. The assay principle is that attachment of the analytes to their corresponding antibodies will result in structural changes of the cell membrane, which is measured as changes occur in the cell membrane electric potential. Measurement and reference electrodes are connected via wiring to a data converter, which assess the membrane potential changes. (**d**) Isolated receptors may also be immobilized inside artificial cell membranes (lipid bilayer). Electrical current flow through this membrane yields information about the analyte. (**e**) CBB based on measurement of the impedance, i.e., the resistance and the capacitance are deduced by measuring the current and voltage across a small empty electrode. Then adherent cells are grown to cover the electrode. As a result, the cell membranes block the current flow causing changes in impedance. (**e**) Cellular functions measured by electric cell substrate impedance sensor (ECIS), which measures various cellular attributes such as: barrier function, cell membrane capacitance, morphology changes, cell motility, cell–cell proximity, proximity of cells to the surface ((**a**), (**b**) and (**d**) are redrawn from Keusgen [168])

sensing is achieved by measuring the changes in the impedance of a small microelectrode in response to AC current flow. High-resolution recording of ECIS are able to indicate changes in cellular morphology at a nanoscale level. This multichambered biosensing system has been used to detect minute pathogen-induced changes in the cellular response that are normally invisible in conventional microscopic techniques. Equipped with greater sensitivity, ECIS also acquires data in real time and is capable of quantifying intracellular and intercellular changes. This sensor has commercial applications in several areas; including in vitro toxicity testing, signal transduction involving GPCRs for drug discovery, cancer cell invasion, and the testing of endothelial cell barrier function [155]. This was one of the initial breakthroughs to bring the concept of a mammalian or higher eukaryotic whole cell-based biosensing system for diagnostics application.

4.8.2 Bioelectric Recognition Assay

Kintzios and colleagues [31] developed a CBB that employed the so-called bioelectric recognition assay (BERA). Animal and plant cells immobilized on BERA sensors are capable of reflecting the electric response to various ligands bound to the cells. This sensor measures the changes to the cell membrane potential due to either cell–analyte interactions or the oxidation of membrane lipids. For example, the positive signals obtained in the BERA system reflects the binding of viruses to surface antibodies, which changes their membrane potential and not the entry of viruses in immobilized cells (Fig. 3c). Further refinement of membrane engineering technology to allow insertion (electroinsertion method) of virus-specific antibodies on the surface of cultured mammalian cells (such as fibroblasts) has improved the specificity of the sensor [156]. This biosensor technology is also available commercially with an intended target market of food manufacturing and clinical diagnostics.

4.8.3 CANARY™: A B-Lymphocyte-Based Biosensor

One of the first commercially available cell-based sensors having application to food, clinical, and environmental diagnostics or in the biosecurity area is a genetically engineered CBB developed at MIT's Lincoln Laboratory, called Cellular Analysis and Notification of Antigen Risks and Yields (CANARY) [135]. In this system, B-lymphocytes were engineered to express bacteria-specific antibodies (immunoglobulin). Simultaneously, a calcium responsive bioluminescent cytoplasmic protein – "aequorin" from a jellyfish (which emits light when intracellular calcium concentration increases) – was also engineered in such a way that when an antigen (e.g., bacterial cell surface protein) binds to the antibodies on the engineered B-cell surface, a downstream signal transduction cascade triggers intracellular calcium flux. As a result, the jellyfish protein in the cytoplasm of the engineered B-cell will almost instantaneously emit light which can be detected with a luminometer [135, 157]. As such, this biosensor reports the presence of an analyte, such as *E. coli* O157:H7, as soon as it binds to the biosensor's receptors.

4.8.4 Cell-Based Sensor Using Ped-2E9 Hybridoma B Lymphocyte

In another type of CBB the foodborne pathogens *Listeria monocytogenes* or *Bacillus cereus* (whole bacterial cells or toxins) were detected using murine hybridoma B-cells (Ped-2E9) [3, 12, 17, 19, 53, 54, 158]. These organisms or toxins can infect and produce detectable cytotoxicity to Ped-2E9 cells; as a result, the hybridoma cells release ALP. The released enzyme can be detected colorimetrically in 1–6 h, which is directly related to the virulence potential of *Listeria* or *Bacillus* spp. [12, 17, 19].

4.8.5 Artificial Cell-Based Sensor

An artificial cell-based sensor was developed to detect a pore-forming hemolysin produced by *Listeria monocytogenes* – listeriolysin O (LLO) by using a nanocomposite material of small unilamellar liposomes containing fluorescent dyes. The liposomes were encapsulated in porous silica using alcohol-free sol–gel synthesis methods. The immobilized liposomes act as a cellular compartment containing the fluorescent dyes. The released dyes as a result of LLO mediated pore formation report the presence of the toxin [159]. The real advantage of this type of synthetic cell-based sensor is that it is composed purely of "nonliving" components, so shelf-life is very long and preservation is simple when compared to mammalian cell-based sensors.

5 Limitations and Drawbacks of CBBs

Cell-based sensors have emerged as a promising approach to address several issues such as functional identification of an analyte [1], screening of a group of molecules or chemical species [5], or in drug discovery and development as well as environmental monitoring [6]. However, there are several issues which may limit extensive application of CBBs such as specificity, reliability and robustness, stability, and shelf-life.

Some CBBs lack specificity and fail to identify the analyte type [60]. For example, different cellular events, such as cytotoxicity/cell death events or cell-signaling pathways may manifest a similar outcome. Membrane active toxins such as hemolysins and cytolysins from different microbes may cause similar damage to membrane structure of mammalian cells, or different adrenergic compounds might elicit similar cAMP or cGMP mediated cellular signaling events. The discriminatory analysis for the specific detection of an analyte may become difficult when only relying on the hallmark of cellular fate.

CBBs may lack robustness because mammalian cells when cultured in vitro are generally fragile and prone to damage even with slight changes in their growth environment. Thus the cellular microenvironment, e.g., media pH or media composition, must be maintained properly during a CBA. The reliability of the assay depends on maintaining a "true" tissue or organ environment, which often requires sophisticated design of platforms. A classical example of such an issue is observed when neuronal networks or neurons are cultured in 3D instead of 2D formats; they tend to provide much better information [5]. Another limitation of using CBBs is receptor desensitization, which can occur during repetitive analysis. Loss of receptor response is a general problem, regardless of whether the biosensor uses the receptor on the intact living cells or on an artificial support [160, 161].

Another drawback of CBBs is lack of stability and prolonged shelf-life, which are considered crucial properties needed for field application. Long-term cryogenic storage of the cells in liquid nitrogen traditionally reduces the viability, and application of dimethyl sulfoxide (DMSO) as a cryo-protectant might also reduce the number of viable cells [162, 163]. Prolonged storage of the mammalian cells in cultured condition requires extensive facilities such as carbon-dioxide, controlled temperature, and humidity which may hinder in-field application of CBBs. Therefore, to overcome these obstacles, efforts are being made to maximize onsite deployability of CBBs by encapsulating cells in 3D protective materials such as collagen [3]. In addition, field-deployable platforms are being made using cell-support components to prolong the self-life of sensor [164].

One of the most desirable goals is to make the sensors affordable to the users. Since mammalian cell culture is expensive, it is imperative to develop low cost CBBs. Therefore, the development of biosensor platforms utilizing bioluminescence or fluorescence-based detection is possible using a simple, hand-held luminometer or spectrofluorometer.

6 Concluding Remarks and Future of CBBs

As is the case with any technology, biosensing using cell-based approaches is presently undergoing developmental phases. There are limitations (discussed in Sect. 5), but concerted efforts are being made to overcome such limitations to make CBBs available for onsite use and for widespread applications. The power of cellular receptors or signaling pathways has been integrated with analyte-specific antibodies on surface of mammalian cells to render specificity [135]. Application of protease inhibitors [165] and cell-cycle inhibitors [166] in growth medium or cells grown under modified growth conditions [17] have shown to extend shelf-life. These strategies may prove to be encouraging in future development in cell-based sensor for onsite application. With the advent of laser scanning imaging devices [3], micro and nano electronics [167], and fluorescence probes, a new paradigm of cell-based sensing is definitely emerging. The capacity of CBBs to provide useful information concerning physiological responses to a variety of potentially biohazardous analytes coupled with innovative advances in detection platforms substantiates them as the leaders in the next generation of functional biosensing.

Acknowledgments Research in the authors' laboratory was supported through a cooperative agreement with the Agricultural Research Service of the US Department of Agriculture (USDA) project number 1935-42000-035, the Center for Food Safety and Engineering at Purdue University, and USDA-NRI (2005-35603-16338). BF is supported by the USDA National Needs Fellowship.

References

1. Pancrazio JJ, Whelan JP, Borkholder DA, Ma W, Stenger DA (1999) Development and application of cell-based biosensors. Ann Biomed Eng 27:697–711
2. Banerjee P, Bhunia AK (2009) Mammalian cell-based biosensors for pathogens and toxins. Trends Biotechnol 27(3):179–188
3. Banerjee P, Lenz D, Robinson JP, Rickus JL, Bhunia AK (2008) A novel and simple cell-based detection system with a collagen-encapsulated B-lymphocyte cell line as a biosensor for rapid detection of pathogens and toxins. Lab Invest 88:196–206
4. Bhunia AK, Banada PP, Banerjee P, Valadez A, Hirleman ED (2007) Light scattering, fiber optic-and cell-based sensors for sensitive detection of foodborne pathogens. J Rapid Methods Autom Microbiol 15:121–145
5. Stenger DA, Gross GW, Keefer EW, Shaffer KM, Andreadis JD, Ma W, Pancrazio JJ (2001) Detection of physiologically active compounds using cell-based biosensors. Trends Biotechnol 19:304–309
6. Ziegler C (2000) Cell-based biosensors. Fresenius' J Anal Chem 366:552–559
7. Rawson DM, Willmer AJ, Turner AP (1989) Whole-cell biosensors for environmental monitoring. Biosensors 4:299–311
8. Deng J, Schoenbach KH, Buescher ES, Hair PS, Fox PM, Beebe SJ (2003) The effects of intense submicrosecond electrical pulses on cells. Biophys J 84:2709–2714
9. Giaever I, Keese CR (1993) A morphological biosensor for mammalian cells. Nature 366:591–592
10. Bhakdi S, Bayley H, Valeva A, Walev I, Walker B, Kehoe M, Palmer M (1996) Staphylococcal alpha-toxin, streptolysin-O, and Escherichia coli hemolysin: prototypes of pore-forming bacterial cytolysins. Arch Microbiol 165:73–79
11. Gilbert RJ (2002) Pore-forming toxins. Cell Mol Life Sci 59:832–844
12. Bhunia AK, Westbrook DG (1998) Alkaline phosphatase release assay to determine cytotoxicity for Listeria species. Lett Appl Microbiol 26:305–310
13. Low MG, Finean JB (1978) Specific release of plasma membrane enzymes by a phosphatidylinositol-specific phospholipase C. Biochim Biophys Acta 508:565–570
14. Moss DW (1994) Release of membrane-bound enzymes from cells and the generation of isoforms. Clin Chim Acta 226:131–142
15. Hardy SP, Lund T, Granum PE (2001) CytK toxin of Bacillus cereus forms pores in planar lipid bilayers and is cytotoxic to intestinal epithelia. FEMS Microbiol Lett 197:47–51
16. Sekiya K, Futaesaku Y (1998) Characterization of the damage to membranes caused by bacterial cytolysins. J Electron Microsc (Tokyo) 47:543–552
17. Banerjee P, Morgan MT, Rickus JL, Ragheb K, Corvalan C, Robinson JP, Bhunia AK (2007) Hybridoma Ped-2E9 cells cultured under modified conditions can sensitively detect *Listeria monocytogenes* and *Bacillus cereus*. Appl Microbiol Biotechnol 73:1423–1434

18. Gray KM, Bhunia AK (2005) Specific detection of cytopathogenic *Listeria monocytogenes* using a two-step method of immunoseparation and cytotoxicity analysis. J Microbiol Methods 60:259–268
19. Gray KM, Banada PP, O'Neal E, Bhunia AK (2005) Rapid Ped-2E9 cell-based cytotoxicity analysis and genotyping of Bacillus species. J Clin Microbiol 43:5865–5872
20. Lee JH, Mitchell RJ, Kim BC, Cullen DC, Gu MB (2005) A cell array biosensor for environmental toxicity analysis. Biosens Bioelectron 21:500–507
21. Hertzberg RP, Pope AJ (2000) High-throughput screening: new technology for the 21st century. Curr Opin Chem Biol 4:445–451
22. Aravanis AM, DeBusschere BD, Chruscinski AJ, Gilchrist KH, Kobilka BK, Kovacs GT (2001) A genetically engineered cell-based biosensor for functional classification of agents. Biosens Bioelectron 16:571–577
23. Bousse L (1996) Whole cell biosensors. Sens Actuators B Chem 34:270–275
24. Rabinowitz P, Gordon Z, Chudnov D, Wilcox M, Odofin L, Liu A, Dein J (2006) Animals as sentinels of bioterrorism agents. Emerg Infect Dis 12:647–652
25. Gubernot DM, Boyer BL, Moses MS (2008) Animal as early detectors of bioevents: veterinary tools and a framework for animal-human integrated zoonotic disease surveillance. Public Health Rep 123:300–315
26. van der Schalie WH, Gardner HS, Bantle JA, De Rosa CT, Finch RA, Reif JS, Reuter RH, Backer LC, Burger J, Folmar LC, Stokes WS (1999) Animals as sentinels of human health hazards of environmental chemicals. Environ Health Perspect 107:309–315
27. Li N, Tourovskaia A, Folch A (2003) Biology on a chip: microfabrication for studying the behavior of cultured cells. Crit Rev Biomed Eng 31:423–488
28. O'Shaughnessy TJ, Pancrazio JJ (2007) Broadband detection of environmental neurotoxicants. Anal Chem 79:8838–8845
29. Slaughter GE, Hobson R (2009) An impedimetric biosensor based on PC 12 cells for the monitoring of exogenous agents. Biosens Bioelectron 24:1153–1158
30. Benderitter M, Vincent-Genod L, Pouget JP, Voisin P (2003) The cell membrane as a biosensor of oxidative stress induced by radiation exposure: a multiparameter investigation. Radiat Res 159:471–483
31. Kintzios S, Bem F, Mangana O, Nomikou K, Markoulatos P, Alexandropoulos N, Fasseas C, Arakelyan V, Petrou AL, Soukouli K, Moschopoulou G, Yialouris C, Simonian A (2004) Study on the mechanism of Bioelectric Recognition Assay: evidence for immobilized cell membrane interactions with viral fragments. Biosens Bioelectron 20:907–916
32. Kintzios S, Makri O, Pistola E, Matakiadis T, Shi HP, Economou A (2004) Scale-up production of puerarin from hairy roots of Pueraria phaseoloides in an airlift bioreactor. Biotechnol Lett 26:1057–1059
33. Yamazaki V, Sirenko O, Schafer RJ, Nguyen L, Gutsmann T, Brade L, Groves JT (2005) Cell membrane array fabrication and assay technology. BMC Biotechnol 5:18
34. Alves ID, Salgado GF, Salamon Z, Brown MF, Tollin G, Hruby VJ (2005) Phosphatidylethanolamine enhances rhodopsin photoactivation and transducin binding in a solid supported lipid bilayer as determined using plasmon-waveguide resonance spectroscopy. Biophys J 88:198–210
35. Minic J, Grosclaude J, Aioun J, Persuy MA, Gorojankina T, Salesse R, Pajot-Augy E, Hou Y, Helali S, Jaffrezic-Renault N, Bessueille F, Errachid A, Gomila G, Ruiz O, Samitier J (2005) Immobilization of native membrane-bound rhodopsin on biosensor surfaces. Biochim Biophys Acta 1724:324–332
36. Vo-Dinh T, Cullum B (2000) Biosensors and biochips: advances in biological and medical diagnostics. Fresenius' J Anal Chem 366:540–551
37. Lehmann M, Riedel K, Adler K, Kunze G (2000) Amperometric measurement of copper ions with a deputy substrate using a novel *Saccharomyces cerevisiae* sensor. Biosens Bioelectron 15:211–219
38. Mattiasson B (1997) Cell-based biosensors for environmental monitoring with special reference to heavy metal analysis. Res Microbiol 148:533

39. Sanders CA, Rodriguez M Jr, Greenbaum E (2001) Stand-off tissue-based biosensors for the detection of chemical warfare agents using photosynthetic fluorescence induction. Biosens Bioelectron 16:439–446
40. Smutok O, Dmytruk K, Gonchar M, Sibirny A, Schuhmann W (2007) Permeabilized cells of flavocytochrome b(2) over-producing recombinant yeast Hansenula polymorpha as biological recognition element in amperometric lactate biosensors. Biosens Bioelectron 23:599–605
41. Immonen N, Karp M (2007) Bioluminescence-based bioassays for rapid detection of nisin in food. Biosens Bioelectron 22:1982–1987
42. Schmidt A, StandfussGabisch C, Bilitewski U (1996) Microbial biosensor for free fatty acids using an oxygen electrode based on thick film technology. Biosens Bioelectron 11:1139–1145
43. Jiang YQ, Xiao LL, Zhao L, Chen X, Wang XR, Wong KY (2006) Optical biosensor for the determination of BOD in seawater. Talanta 70:97–103
44. Lin L, Xiao LL, Huang S, Zhao L, Cui JS, Wang XH, Chen X (2006) Novel BOD optical fiber biosensor based on co-immobilized microorganisms in ormosils matrix. Biosens Bioelectron 21:1703–1709
45. Neufeld T, Biran D, Popovtzer R, Erez T, Ron EZ, Rishpon J (2006) Genetically engineered pfabA pfabR bacteria: an electrochemical whole cell biosensor for detection of water toxicity. Anal Chem 78:4952–4956
46. Sakaguchi T, Kitagawa K, Ando T, Murakami Y, Morita Y, Yamamura A, Yokoyama K, Tamiya E (2003) A rapid BOD sensing system using luminescent recombinants of Escherichia coli. Biosens Bioelectron 19:115–121
47. Sakaguchi T, Morioka Y, Yamasaki M, Iwanaga J, Beppu K, Maeda H, Morita Y, Tamiya E (2007) Rapid and onsite BOD sensing system using luminous bacterial cells-immobilized chip. Biosens Bioelectron 22:1345–1350
48. Chambers J, Ames RS, Bergsma D, Muir A, Fitzgerald LR, Hervieu G, Dytko GM, Foley JJ, Martin J, Liu WS, Park J, Ellis C, Ganguly S, Konchar S, Cluderay J, Leslie R, Wilson S, Sarau HM (1999) Melanin-concentrating hormone is the cognate ligand for the orphan G-protein-coupled receptor SLC-1. Nature 400:261–265
49. Pietrangelo A (2002) Mechanism of iron toxicity. Adv Exp Med Biol 509:19–43
50. Pietrangelo A, Montosi G, Garuti C, Contri M, Giovannini F, Ceccarelli D, Masini A (2002) Iron-induced oxidant stress in nonparenchymal liver cells: mitochondrial derangement and fibrosis in acutely iron-dosed gerbils and its prevention by silybin. J Bioenerg Biomembr 34:67–79
51. Rudolph AS, Reasor J (2001) Cell and tissue based technologies for environmental detection and medical diagnostics. Biosens Bioelectron 16:429–431
52. Sacco MG, Amicone L, Cato EM, Filippini D, Vezzoni P, Tripodi M (2004) Cell-based assay for the detection of chemically induced cellular stress by immortalized untransformed transgenic hepatocytes. BMC Biotechnol 4:5
53. Bhunia AK, Steele PJ, Westbrook DG, Bly LA, Maloney TP, Johnson MG (1994) A six-hour in vitro virulence assay for *Listeria monocytogenes* using myeloma and hybridoma cells from murine and human sources. Microb Pathog 16:99–110
54. Bhunia AK, Westbrook DG, Story R, Johnson MG (1995) Frozen stored murine hybridoma cells can be used to determine the virulence of *Listeria monocytogenes*. J Clin Microbiol 33:3349–3351
55. Meister M, Pine J, Baylor DA (1994) Multi-neuronal signals from the retina: acquisition and analysis. J Neurosci Methods 51:95–106
56. Segev R, Goodhouse J, Puchalla J, Berry MJ 2nd (2004) Recording spikes from a large fraction of the ganglion cells in a retinal patch. Nat Neurosci 7:1154–1161
57. Hafner F (2000) Cytosensor Microphysiometer: technology and recent applications. Biosens Bioelectron 15:149–158
58. Wang P, Xu GX, Qin LF, Xu Y, Li Y, Li R (2005) Cell-based biosensors and its application in biomedicine. Sens Actuators B Chem 108:576–584

59. Liu Q, Cai H, Xu Y, Xiao L, Yang M, Wang P (2007) Detection of heavy metal toxicity using cardiac cell-based biosensor. Biosens Bioelectron 22:3224–3229
60. Gilchrist KH, Giovangrandi L, Whittington RH, Kovacs GT (2005) Sensitivity of cell-based biosensors to environmental variables. Biosens Bioelectron 20:1397–1406
61. Whittington RH, Chen MQ, Giovangrandi L, Kovacs GA (2006) Temporal resolution of stimulation threshold: a tool for electrophysiologic analysis. Conf Proc IEEE Eng Med Biol Soc 1:3891–3894
62. Gilchrist KH (2003) Characterization and validation of cell-based biosensors. PhD dissertation, Stanford University, United States–California Retrieved January 13, 2008, from ProQuest Digital Dissertations database (Publication No. AAT 3104228), pp 3–6
63. Keefer EW, Gramowski A, Stenger DA, Pancrazio JJ, Gross GW (2001) Characterization of acute neurotoxic effects of trimethylolpropane phosphate via neuronal network biosensors. Biosens Bioelectron 16:513–525
64. Pancrazio JJ, Keefer EW, Ma W, Stenger DA, Gross GW (2001) Neurophysiologic effects of chemical agent hydrolysis products on cortical neurons in vitro. Neurotoxicology 22:393–400
65. Pancrazio JJ, Gray SA, Shubin YS, Kulagina N, Cuttino DS, Shaffer KM, Eisemann K, Curran A, Zim B, Gross GW, O'Shaughnessy TJ (2003) A portable microelectrode array recording system incorporating cultured neuronal networks for neurotoxin detection. Biosens Bioelectron 18:1339–1347
66. Selinger JV, Pancrazio JJ, Gross GW (2004) Measuring synchronization in neuronal networks for biosensor applications. Biosens Bioelectron 19:675–683
67. Bhunia AK (2008) Biosensors and bio-based methods for the separation and detection of foodborne pathogens. In: Taylor S (ed) Advances in food and nutrition research. Elsevier, San Diego, Vol 54, pp 1–44
68. Ehret R, Baumann W, Brischwein M, Schwinde A, Stegbauer K, Wolf B (1997) Monitoring of cellular behaviour by impedance measurements on interdigitated electrode structures. Biosens Bioelectron 12:29–41
69. Ehret R, Baumann W, Brischwein M, Schwinde A, Wolf B (1998) On-line control of cellular adhesion with impedance measurements using interdigitated electrode structures. Med Biol Eng Comput 36:365–370
70. Tiruppathi C, Malik AB, Del Vecchio PJ, Keese CR, Giaever I (1992) Electrical method for detection of endothelial cell shape change in real time: assessment of endothelial barrier function. Proc Natl Acad Sci USA 89:7919–7923
71. Kowolenko M, Keese CR, Lawrence DA, Giaever I (1990) Measurement of macrophage adherence and spreading with weak electric fields. J Immunol Methods 127:71–77
72. Borkholder DA, Bao J, Maluf NI, Perl ER, Kovacs GT (1997) Microelectrode arrays for stimulation of neural slice preparations. J Neurosci Methods 77:61–66
73. Pancrazio JJ, Bey PP Jr, Loloee A, Manne S, Chao HC, Howard LL, Gosney WM, Borkholder DA, Kovacs GT, Manos P, Cuttino DS, Stenger DA (1998) Description and demonstration of a CMOS amplifier-based-system with measurement and stimulation capability for bioelectrical signal transduction. Biosens Bioelectron 13:971–979
74. Lebrun M, Mengaud J, Ohayon H, Nato F, Cossart P (1996) Internalin must be on the bacterial surface to mediate entry of Listeria monocytogenes into epithelial cells. Mol Microbiol 21:579–592
75. Lundstrom I, Svensson S (1998) Biosensing with G-protein coupled receptor systems. Biosens Bioelectron 13:689–695
76. Subrahmanyam S, Piletsky SA, Turner AP (2002) Application of natural receptors in sensors and assays. Anal Chem 74:3942–3951
77. Wijesuriya DC, Rechnitz GA (1993) Biosensors based on plant and animal tissues. Biosens Bioelectron 8:155–160
78. Golden RJ, Noller KL, Titus-Ernstoff L, Kaufman RH, Mittendorf R, Stillman R, Reese EA (1998) Environmental endocrine modulators and human health: an assessment of the biological evidence. Crit Rev Toxicol 28:109–227

79. Haga T (1995) Receptor biochemistry. In: Meyers RA (ed) Molecular biology and biotechnology, a comprehensive desk reference. VCH, New York
80. Connolly CN, Wafford KA (2004) The Cys-loop superfamily of ligand-gated ion channels: the impact of receptor structure on function. Biochem Soc Trans 32:529–534
81. Whitaker RD, Walt DR (2007) Multianalyte single-cell analysis with multiple cell lines using a fiber-optic array. Anal Chem 79:9045–9053
82. Robinson DR, Wu YM, Lin SF (2000) The protein tyrosine kinase family of the human genome. Oncogene 19:5548–5557
83. Kaczmarski RS, Mufti GJ (1991) The cytokine receptor superfamily. Blood Rev 5:193–203
84. Milev P, Monnerie H, Popp S, Margolis RK, Margolis RU (1998) The core protein of the chondroitin sulfate proteoglycan phosphacan is a high-affinity ligand of fibroblast growth factor-2 and potentiates its mitogenic activity. J Biol Chem 273:21439–21442
85. Hauck CR, Agerer F, Muenzner P, Schmitter T (2006) Cellular adhesion molecules as targets for bacterial infection. Eur J Cell Biol 85:235–242
86. Weiss AA, Iyer SS (2007) Glycomics aims to interpret the third molecular language of cells. Microbe 2:489–497
87. Hauck CR (2002) Cell adhesion receptors – signaling capacity and exploitation by bacterial pathogens. Med Microbiol Immunol 191:55–62
88. Patti JM, Allen BL, McGavin MJ, Hook M (1994) MSCRAMM-mediated adherence of microorganisms to host tissues. Annu Rev Microbiol 48:585–617
89. Schwarz-Linek U, Werner JM, Pickford AR, Gurusiddappa S, Kim JH, Pilka ES, Briggs JA, Gough TS, Hook M, Campbell ID, Potts JR (2003) Pathogenic bacteria attach to human fibronectin through a tandem beta-zipper. Nature 423:177–181
90. Frankel G, Lider O, Hershkoviz R, Mould AP, Kachalsky SG, Candy DC, Cahalon L, Humphries MJ, Dougan G (1996) The cell-binding domain of intimin from enteropathogenic Escherichia coli binds to beta1 integrins. J Biol Chem 271:20359–20364
91. Kenny B, DeVinney R, Stein M, Reinscheid DJ, Frey EA, Finlay BB (1997) Enteropathogenic E. coli (EPEC) transfers its receptor for intimate adherence into mammalian cells. Cell 91:511–520
92. Kenny B (1999) Phosphorylation of tyrosine 474 of the enteropathogenic Escherichia coli (EPEC) Tir receptor molecule is essential for actin nucleating activity and is preceded by additional host modifications. Mol Microbiol 31:1229–1241
93. Hauck CR, Meyer TF (2003) 'Small' talk: Opa proteins as mediators of Neisseria-host-cell communication. Curr Opin Microbiol 6:43–49
94. Steinberg MS, McNutt PM (1999) Cadherins and their connections: adhesion junctions have broader functions. Curr Opin Cell Biol 11:554–560
95. Pertz O, Bozic D, Koch AW, Fauser C, Brancaccio A, Engel J (1999) A new crystal structure, Ca2+ dependence and mutational analysis reveal molecular details of E-cadherin homoassociation. EMBO J 18:1738–1747
96. Bierne H, Sabet C, Personnic N, Cossart P (2007) Internalins: a complex family of leucine-rich repeat-containing proteins in *Listeria monocytogenes*. Microbes Infect 9:1156–1166
97. Braun L, Ghebrehiwet B, Cossart P (2000) gC1q-R/p32, a C1q-binding protein, is a receptor for the InlB invasion protein of *Listeria monocytogenes*. EMBO J 19:1458–1466
98. Shen Y, Naujokas M, Park M, Ireton K (2000) InIB-dependent internalization of Listeria is mediated by the Met receptor tyrosine kinase. Cell 103:501–510
99. Gaillard JL, Berche P, Frehel C, Gouin E, Cossart P (1991) Entry of *L. monocytogenes* into cells is mediated by internalin, a repeat protein reminiscent of surface antigens from Gram-positive cocci. Cell 65:1127–1141
100. Mengaud J, Ohayon H, Gounon P, Mege RM, Cossart P (1996) E-cadherin is the receptor for internalin, a surface protein required for entry of *L. monocytogenes* into epithelial cells. Cell 84:923–932
101. Dramsi S, Cossart P (1998) Intracellular pathogens and the actin cytoskeleton. Annu Rev Cell Dev Biol 14:137–166

102. Lasa I, Cossart P (1996) Actin-based bacterial motility: towards a definition of the minimal requirements. Trends Cell Biol 6:109–114
103. Pandiripally VK, Westbrook DG, Sunki GR, Bhunia AK (1999) Surface protein p104 is involved in adhesion of *Listeria monocytogenes* to human intestinal cell line, Caco-2. J Med Microbiol 48:117–124
104. Santiago NI, Zipf A, Bhunia AK (1999) Influence of temperature and growth phase on expression of a 104-kilodalton *Listeria* adhesion protein in *Listeria monocytogenes*. Appl Environ Microbiol 65:2765–2769
105. Wampler JL, Kim KP, Jaradat Z, Bhunia AK (2004) Heat shock protein 60 acts as a receptor for the Listeria adhesion protein in Caco-2 cells. Infect Immun 72:931–936
106. Jaradat ZW, Bhunia AK (2003) Adhesion, invasion, and translocation characteristics of Listeria monocytogenes serotypes in Caco-2 cell and mouse models. Appl Environ Microbiol 69:3640–3645
107. Kim KP, Jagadeesan B, Burkholder KM, Jaradat ZW, Wampler JL, Lathrop AA, Morgan MT, Bhunia AK (2006) Adhesion characteristics of Listeria adhesion protein (LAP)-expressing *Escherichia coli* to Caco-2 cells and of recombinant LAP to eukaryotic receptor Hsp60 as examined in a surface plasmon resonance sensor. FEMS Microbiol Lett 256:324–332
108. Campadelli-Fiume G (2000) Virus receptor arrays, CD46 and human herpesvirus 6. Trends Microbiol 8:436–438
109. Gruenheid S, Gatzke L, Meadows H, Tufaro F (1993) Herpes simplex virus infection and propagation in a mouse L cell mutant lacking heparan sulfate proteoglycans. J Virol 67:93–100
110. Ugolini S, Mondor I, Sattentau QJ (1999) HIV-1 attachment: another look. Trends Microbiol 7:144–149
111. Spear PG, Shieh MT, Herold BC, WuDunn D, Koshy TI (1992) Heparan sulfate glycosaminoglycans as primary cell surface receptors for herpes simplex virus. Adv Exp Med Biol 313:341–353
112. Berger EA, Murphy PM, Farber JM (1999) Chemokine receptors as HIV-1 coreceptors: roles in viral entry, tropism, and disease. Annu Rev Immunol 17:657–700
113. Schroeder K, Neagle B (1996) FLIPR: a new instrument for accurate, high throughput optical screening. J Biomol Screen 1:75
114. Gonzalez JE, Oades K, Leychkis Y, Harootunian A, Negulescu PA (1999) Cell-based assays and instrumentation for screening ion-channel targets. Drug Discov Today 4:431–439
115. Kiss L, Bennett PB, Uebele VN, Koblan KS, Kane SA, Neagle B, Schroeder K (2003) High throughput ion-channel pharmacology: planar-array-based voltage clamp. Assay Drug Dev Technol 1:127–135
116. Zuck P, Lao Z, Skwish S, Glickman JF, Yang K, Burbaum J, Inglese J (1999) Ligand-receptor binding measured by laser-scanning imaging. Proc Natl Acad Sci USA 96: 11122–11127
117. Lee JY, Miraglia S, Yan X, Swartzman E, Cornell-Kennon S, Mellentin-Michelotti J, Bruseo C, France DS (2003) Oncology drug discovery applications using the FMAT 8100 HTS system. J Biomol Screen 8:81–88
118. Conway BR, Minor LK, Xu JZ, Gunnet JW, DeBiasio R, D'Andrea MR, Rubin R, DeBiasio R, Giuliano K, Zhou LB, Demarest KT (1999) Quantification of G-protein coupled receptor internalization using G-protein coupled receptor-green fluorescent protein conjugates with the ArrayScan (TM) high-content screening system. J Biomol Screen 4:75–86
119. Gasparri F, Mariani M, Sola F, Galvani A (2004) Quantification of the proliferation index of human dermal fibroblast cultures with the ArrayScan high-content screening reader. J Biomol Screen 9:232–243
120. Trask OJ Jr, Baker A, Williams RG, Nickischer D, Kandasamy R, Laethem C, Johnston PA, Johnston PA (2006) Assay development and case history of a 32K-biased library high-content MK2-EGFP translocation screen to identify p38 mitogen-activated protein kinase inhibitors on the ArrayScan 3.1 imaging platform. Methods Enzymol 414:419–439

121. Williams RG, Kandasamy R, Nickischer D, Trask OJ Jr, Laethem C, Johnston PA, Johnston PA (2006) Generation and characterization of a stable MK2-EGFP cell line and subsequent development of a high-content imaging assay on the Cellomics ArrayScan platform to screen for p38 mitogen-activated protein kinase inhibitors. Methods Enzymol 414:364–389
122. Liu Q, Huang H, Cai H, Xu Y, Li Y, Li R, Wang P (2007) Embryonic stem cells as a novel cell source of cell-based biosensors. Biosens Bioelectron 22:810–815
123. May KML, Wang Y, Bachas LG, Anderson KW (2004) Development of a whole-cell-based biosensor for detecting histamine as a model toxin. Anal Chem 76:4156–4161
124. Davila JC, Cezar GG, Thiede M, Strom S, Miki T, Trosko J (2004) Use and application of stem cells in toxicology. Toxicol Sci 79:214–223
125. Raucy JL, Mueller L, Duan K, Allen SW, Strom S, Lasker JM (2002) Expression and induction of CYP2C P450 enzymes in primary cultures of human hepatocytes. J Pharmacol Exp Ther 302:475–482
126. Lavon N, Benvenisty N (2003) Differentiation and genetic manipulation of human embryonic stem cells and the analysis of the cardiovascular system. Trends Cardiovasc Med 13:47–52
127. Rambhatla L, Chiu CP, Kundu P, Peng Y, Carpenter MK (2003) Generation of hepatocyte-like cells from human embryonic stem cells. Cell Transplant 12:1–11
128. Trosko JE (2003) The role of stem cells and gap junctional intercellular communication in carcinogenesis. J Biochem Mol Biol 36:43–48
129. Hanson GT, Hanson BJ (2008) Fluorescent probes for cellular assays. Comb Chem High Throughput Screen 11:505–513
130. Wang H-Y, Bao N, Lu C (2008) A microfluidic cell array with individually addressable culture chambers. Biosens Bioelectron 24:613–617
131. Lee RM, Choi H, Shin J-S, Kim K, Yoo K-H (2009) Distinguishing between apoptosis and necrosis using a capacitance sensor. Biosens Bioelectron 24(8):2586–2591
132. Tong C, Shi B, Xiao X, Liao H, Zheng Y, Shen G, Tang D, Liu X (2009) An annexin V based biosensor for quantitatively detecting early apoptotic cells. Biosens Bioelectron 24 (6):1777–1782
133. Gil GC, Mitchell RJ, Chang ST, Gu MB (2000) A biosensor for the detection of gas toxicity using a recombinant bioluminescent bacterium. Biosens Bioelectron 15:23–30
134. Hay AG, Rice JF, Applegate BM, Bright NG, Sayler GS (2000) A bioluminescent whole-cell reporter for detection of 2,4-dichlorophenoxyacetic acid and 2,4-dichlorophenol in soil. Appl Environ Microbiol 66:4589–4594
135. Rider TH, Petrovick MS, Nargi FE, Harper JD, Schwoebel ED, Mathews RH, Blanchard DJ, Bortolin LT, Young AM, Chen J, Hollis MA (2003) A B cell-based sensor for rapid identification of pathogens. Science 301:213–215
136. Shingleton JT, Applegate BA, Baker AJ, Sayler GS, Bienkowski PR (2001) Quantification of toluene dioxygenase induction and kinetic modeling of TCE cometabolism by Pseudomonas putida TVA8. Biotechnol Bioeng 76:341–350
137. O'Riordan TC, Buckley D, Ogurtsov V, O'Connor R, Papkovsky DB (2000) A cell viability assay based on monitoring respiration by optical oxygen sensing. Anal Biochem 278:221–227
138. Sturzl M, Konrad A, Sander G, Wies E, Neipel F, Naschberger E, Reipschlager S, Gonin-Laurent N, Horch RE, Kneser U, Hohenberger W, Erfle H, Thurau M (2008) High throughput screening of gene functions in mammalian cells using reversely transfected cell arrays: review and protocol. Comb Chem High Throughput Screen 11:159–172
139. Durick K, Negulescu P (2001) Cellular biosensors for drug discovery. Biosens Bioelectron 16:587–592
140. Kumar HS, Karunasagar I, Teizou T, Shima K, Yamasaki S (2004) Characterisation of Shiga toxin-producing *Escherichia coli* (STEC) isolated from seafood and beef. FEMS Microbiol Lett 233:173–178

141. Noda M, Yutsudo T, Nakabayashi N, Hirayama T, Takeda Y (1987) Purification and some properties of Shiga-like toxin from *Escherichia coli* 0157:H7 that is immunologically identical to Shiga toxin. Microb Pathog 2:339–349
142. Picot L, Chevalier S, Mezghani-Abdelmoula S, Merieau A, Lesouhaitier O, Leroux P, Cazin L, Orange N, Feuilloley MG (2003) Cytotoxic effects of the lipopolysaccharide from *Pseudomonas fluorescens* on neurons and glial cells. Microb Pathog 35:95–106
143. Ngamwongsatit P, Banada PP, Panbangred W, Bhunia AK (2008) WST-1-based cell cytotoxicity assay as a substitute for MTT-based assay for rapid detection of toxigenic *Bacillus species* using CHO cell line. J Microbiol Methods 73:211–215
144. Sakurazawa T, Ohkusa T (2005) Cytotoxicity of organic acids produced by anaerobic intestinal bacteria on cultured epithelial cells. J Gastroenterol 40:600–609
145. Saliba AM, de Assis MC, Nishi R, Raymond B, Marques Ede A, Lopes UG, Touqui L, Plotkowski MC (2006) Implications of oxidative stress in the cytotoxicity of *Pseudomonas aeruginosa* ExoU. Microbes Infect 8:450–459
146. Lee J, Cuddihy MJ, Kotov NA (2008) Three-dimensional cell culture matrices: state of the art. Tissue Eng Part B Rev 14:61–86
147. Liu J, Kuznetsova LA, Edwards GO, Xu J, Ma M, Purcell WM, Jackson SK, Coakley WT (2007) Functional three-dimensional HepG2 aggregate cultures generated from an ultrasound trap: comparison with HepG2 spheroids. J Cell Biochem 102:1180–1189
148. Pampaloni F, Reynaud EG, Stelzer EHK (2007) The third dimension bridges the gap between cell culture and live tissue. Nat Rev Mol Cell Biol 8:839–845
149. Yamada KM, Cukierman E (2007) Modeling tissue morphogenesis and cancer in 3D. Cell 130:601–610
150. Nickerson C, Richter E, Ott C (2007) Studying host–pathogen interactions in 3-D: organotypic models for infectious disease and drug development. J Neuroimmune Pharmacol 2:26–31
151. Smith YC, Grande KK, Rasmussen SB, O'Brien AD (2006) Novel three-dimensional organoid model for evaluation of the interaction of uropathogenic *Escherichia coli* with terminally differentiated human urothelial cells. Infect Immun 74:750–757
152. Curtis T, Naal RMZG, Batt C, Tabb J, Holowka D (2008) Development of a mast cell-based biosensor. Biosens Bioelectron 23:1024–1031
153. Straub TM, KHz B, Orosz-Coghlan P, Dohnalkova A, Mayer BK, Bartholomew RA, Valdez CO, Bruckner-Lea CJ, Gerba CP, Abbaszadegan M, Nickerson CA (2007) In vitro cell culture infectivity assay for human Noroviruses. Emerg Infect Dis 13:396–403
154. Lee M-Y, Kumar RA, Sukumaran SM, Hogg MG, Clark DS, Dordick JS (2008) Three-dimensional cellular microarray for high-throughput toxicology assays. Proc Nat Acad Sci USA 105:59–63
155. Campbell CE, Laane MM, Haugarvoll E, Giaever I (2007) Monitoring viral-induced cell death using electric cell-substrate impedance sensing. Biosens Bioelectron 23:536–542
156. Moschopoulou G, Vitsa K, Bem F, Vassilakos N, Perdikaris A, Blouhos P, Yialouris C, Frosyniotis D, Anthopoulos I, Mangana O, Nomikou K, Rodeva V, Kostova D, Grozeva S, Michaelides A, Simonian A, Kintzios S (2008) Engineering of the membrane of fibroblast cells with virus-specific antibodies: a novel biosensor tool for virus detection. Biosens Bioelectron 24:1033–1036
157. Relman DA (2003) Shedding light on microbial detection. N Engl J Med 349:2162–2163
158. Shroyer ML, Bhunia AK (2003) Development of a rapid 1-h fluorescence-based cytotoxicity assay for Listeria species. J Microbiol Methods 55:35–40
159. Zhao J, Jedlicka SS, Lannu JD, Bhunia AK, Rickus JL (2006) Liposome-doped nanocomposites as artificial cell-based biosensors: detection of listeriolysin O. Biotechnol Prog 22:32–37
160. Barak LS, Salahpour A, Zhang X, Masri B, Sotnikova TD, Ramsey AJ, Violin JD, Lefkowitz RJ, Caron MG, Gainetdinov RR (2008) Pharmacological characterization of membrane-expressed human trace amine-associated receptor 1 (TAAR1) by a bioluminescence resonance energy transfer cAMP biosensor. Mol Pharmacol 74:585–594

161. Rogers KR, Valdes JJ, Eldefrawi ME (1991) Effects of receptor concentration, media pH and storage on nicotinic receptor-transmitted signal in a fiber-optic biosensor. Biosens Bioelectron 6:1–8
162. Hanslick JL, Lau K, Noguchi KK, Olney JW, Zorumski CF, Mennerick S, Farber NB (2009) Dimethyl sulfoxide (DMSO) produces widespread apoptosis in the developing central nervous system. Neurobiol Dis 34(1):1–10
163. Korbutt GS, Rayat GR, Ezekowitz J, Rajotte RV (1997) Cryopreservation of rat pancreatic islets: effect of ethylene glycol on islet function and cellular composition. Transplantation 64:1065–1070
164. Gilchrist KH, Barker VN, Fletcher LE, DeBusschere BD, Ghanouni P, Giovangrandi L, Kovacs GT (2001) General purpose, field-portable cell-based biosensor platform. Biosens Bioelectron 16:557–564
165. Baust JM, Van B, Baust JG (2000) Cell viability improves following inhibition of cryopreservation-induced apoptosis. In Vitro Cell Dev Biol Anim 36:262–270
166. Nakagawa T, Yamaguchi M (2005) Overexpression of regucalcin suppresses apoptotic cell death in cloned normal rat kidney proximal tubular epithelial NRK52E cells: change in apoptosis-related gene expression. J Cell Biochem 96:1274–1285
167. Kovacs GT (2003) Electronic sensors with living cellular components. Proc IEEE 91:915–929
168. Keusgen M (2002) Biosensors: new approaches in drug discovery. Naturwissenschaften 89:433–444
169. Burrell GA, Seibert FM (1916) Gases found in coal mines. Miners' Circ 14
170. Schwabe CW (1984) Animals as monitors of the environment. Veterinary Medicine and Human Health, 3rd ed. Williams and Wilkins, Baltimore, MD, USA, pp 562–578
171. Veterinarian (1874) The effects of the fog on cattle in London. Veterinarian 47:1–4
172. Veterinarian (1874) The effects of the recent fog on the Smithfield Show and the London dairies. Veterinarian 47:32–33
173. Haring CM, Meyer KF (1915) Investigations of livestock conditions with horses in the Selby smoke zone. Calif Hurrau Mines Bull 98
174. Holm LW, Wheat JD, Rhode EA, Firch G (1953) Treatment of chronic lead poisoning in horses with calcium disodium ethylenediaminetetreacetate. J Am Vet Assoc 123:383–388
175. Kurland LT, Faro SN, Siedler H (1960) Minamata disease. World Neurol 1:370–395
176. Kuratsune M, Yoshimura T, Matsuzaka J, Yamaguchi A (1972) Epidemiologic study on Yusho, a poisoning caused by ingestion of rice oil contaminated with a commercial brand of polychlorinated biphenyls. Environ Health Perspect 1:119–128
177. Van Kampen KR, James LF, Rasmussen J, Huffaker RH, Fawcett MO (1969) Organic phosphate poisoning of sheep in Skull Valley, Utah. J Am Vet Med Asso 154:623–630
178. Case AA, Coffman JR (1973) Waste oil: toxic for horses. Vet Clin North Am 3:273–277
179. Carter CD, Kimbrough RD, Liddle JA (1975) Tetrachlorodibenzodioxin: an accidental poisoning episode in horse arenas. Science 188:738–740
180. Jackson TF, Halbert FL (1974) A toxic syndrome associated with the feeding of polybrominated biphenyl-contaminated protein concentrate to dairy cattle. J Am Vet Med Asso 165:437–439
181. Welborn JA, Allen R, Byker G, DeGrow S, Hertel J, Noordhoek R, Koons D (1975) The contamination crisis in Michigan: polybrominated biphenyls. Senate Special Investigating Committee, Lansing, MI
182. Guillette LJ Jr, Gross TS, Masson GR, Matter JM, Percival HF, Woodward AR (1994) Developmental abnormalities of the gonad and abnormal sex hormone concentrations in juvenile alligators from contaminated and control lakes in Florida. Environ Health Perspect 102:680–688

Adv Biochem Engin/Biotechnol (2010) 117: 57–75
DOI: 10.1007/10_2009_22

Published online: 21 January 2010

Fluorescent and Bioluminescent Cell-Based Sensors: Strategies for Their Preservation

Amol Date, Patrizia Pasini, and Sylvia Daunert

Abstract Luminescent whole-cell biosensing systems have been developed for a variety of analytes of environmental, clinical, and biological interest. These analytical tools allow for sensitive, rapid, simple, and inexpensive quantitative detection of target analytes. Furthermore, they can be designed to be nonspecific, semispecific, or highly specific/selective. A notable feature of such sensing systems employing living cells is that they provide information on the analyte bioavailability and activity. These characteristics, along with their suitability to miniaturization, make cell-based sensors ideal for field applications. However, a major limitation to on-site use is their "shelf-life." To address this problem, various methods for preservation of sensing cells have been reported, including freeze-drying, immobilization in different types of matrices, and formation of spores. Among these, the use of spores emerged as a promising strategy for long-term storage of whole-cell sensing systems at room temperature as well as in extreme environmental conditions.

Keywords Bioluminescence • Cell preservation • Fluorescence • Spores • Whole-cell biosensing systems

Contents

A. Date, P. Pasini, and S. Daunert (✉)
Department of Chemistry, University of Kentucky, Lexington, KY 40506-0055, USA
e-mail: daunert@uky.edu

1 Introduction

A biosensor is an analytical device comprised of a biological sensing component coupled to a transduction element that produces a measurable signal in response to an environmental change or a target analyte. Various biological recognition elements such as binding proteins, antibodies, enzymes, and whole cells, among others, have been employed in biosensors. In particular, the use of microbial cells as bioreporters for detection of environmental pollutants and other biologically relevant chemicals has become a trend in environmental, clinical, and biological analysis. Whole-cell bacterial biosensing systems have been developed by inserting a plasmid into the cell that encodes for a regulatory protein and a reporter protein under the control of an inducible operator/promoter. The reporter protein is under the promoter's transcriptional control and is expressed in a dose-dependent fashion in the presence of specific compounds/analytes recognized by the regulatory protein, allowing for quantitative detection of the analytes. Specifically, fluorescent and bioluminescent proteins such as the green fluorescent protein (GFP) and its variants, red fluorescent protein (DsRed), bacterial, firefly, and sea pansy luciferases, as well as β-galactosidase, that can be detected by chemiluminescent substrates, have been used as reporter proteins in whole-cell sensing. Fluorescence and bioluminescence detection provides high sensitivity due to the high quantum yields of luminescence reactions [1], which results in a linear relationship between the analyte concentration and the light signal occurring over a wide dynamic range of concentrations. The main characteristics responsible for the widespread application of fluorescent and bioluminescent reporter genes in whole-cell sensing are their low cost, safety, rapid response and convenience of use. Both types of reporter genes allow in situ analysis of biological samples. Fluorescent reporter proteins such as the GFP and its variants are autofluorescent and, therefore, do not require addition of substrates or cofactors. The stability of fluorescent proteins at biological pH and the lack of endogenous homologues make them an attractive tool for detection of a variety of analytes in many different assay configurations and platforms. However, the overall sensitivity of such proteins may be hindered due to the background fluorescence of the biological components in the analyzed samples. In that regard, bioluminescent reporter proteins such as bacterial, firefly, and sea pansy luciferases have a distinct advantage over fluorescent proteins in that they can deliver higher assay sensitivity. This is because bioluminescent proteins do not require an external excitation source, thus eliminating the possible interference by fluorescent compounds present in biological systems. A drawback of the use of bioluminescent reporters relates to the requirement for substrates and/or cofactors to trigger the bioluminescence reaction. In the case of whole-cell biosensors that employ intact cells, this may pose a problem given that the substrates need to be able to cross the cell membrane, and thus permeability issues may arise. When bacterial luciferase is employed as the reporter, the need for exogenous substrate is eliminated if the entire *luxCDABE* gene cassette is used. In this cassette, the *luxA* and *luxB* genes code for two distinct bacterial luciferase subunits, while the other genes code for enzymes involved in the synthesis of the substrate, namely, a long-chain aldehyde [2]. It should also be noted that some bioluminescent proteins such as certain bacterial luciferases are heat sensitive, a factor that restricts their widespread use. A versatile and widely used reporter in a variety of applications is β-galactosidase. The versatility of this reporter protein stems from the fact that its enzymatic activity can be measured by electrochemical, fluorescent, chemiluminescent, and colorimetric detection methods. From all these detection methods, the one that affords better detection limits when using β-galactosidase is chemiluminescence [3].

Bacterial sensing systems include two principal types, based on the expression of the reporter protein, which can be either constitutive or inducible. In the former type the reporter protein is constantly expressed by placing the reporter gene under the control of a constitutively active promoter. The decrease in intensity of the signal produced by the reporter protein indicates a decrease in metabolic activity of the cells, which is due to any toxicants present in the environment of the cells (Fig. 1). Use of constitutive systems to monitor the toxicity of aquatic samples was first reported almost three decades ago [4]. One of the prominent features of these nonspecific bacterial biosensors is that they can be used to detect mixed toxicants [5]. Unpredictable additive effects of such mixed toxicants in complex mixtures and environmental samples can also be assessed [6]. Furthermore, commercialized testing kits that make use of constitutive expression reporter cells to monitor the toxicity of samples are now available. Specifically, testing kits such as Microtox and Lumistox are based on inhibition of luminescence from the spontaneously bioluminescent bacterium *Vibrio fischeri*, caused by the presence of toxic compounds. While constitutive systems are effective in providing information about the toxicity of complex mixtures and samples, there are some challenges that still need to be addressed. Complex environment factors such as distribution of nutrients and inhibitory compounds that inactivate expression of the reporter gene may yield false positive results [7]. Moreover, decrease in the metabolic activity of the cells

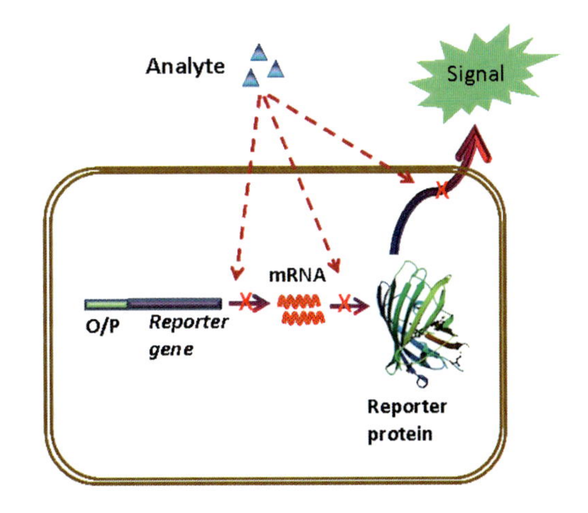

Fig. 1 Schematic representation of a constitutive whole-cell sensing system in which the reporter protein is placed under the transcriptional control of a constitutively active promoter. In the absence of toxic chemicals the reporter protein is constantly expressed and a steady signal is measured. In the presence of toxic chemicals the signal is reduced or totally inhibited

for any reasons other than the presence of toxic chemicals could yield unreliable results. In addition, potassium, sodium, and magnesium ions have been shown to alter the luminescence in *V. fischeri* [8].

Alternatively, in inducible bacterial sensing systems a measurable signal is elicited only upon activation by target analytes or certain stress factors. The sensor's response consists of a dose-dependent increase in the production of reporter protein and, consequently, in the signal intensity. In the case of sensing systems responding to target compounds (Fig. 2), the presence of a biorecognition element, which is comprised of a regulatory/recognition protein and a specific promoter region of DNA, ensures that the system is extremely specific/selective to those molecules or classes of molecules. For example, regulatory sequences from heavy metal resistant bacteria have been fused to reporter genes for construction of highly specific and sensitive bacterial biosensors for heavy metal detection. Inducible whole-cell biosensors for stress factors are constructed by placing the reporter gene under the control of the promoter of a stress-response regulon. Such promoters are induced by various stimuli/chemicals; therefore, these biosensors are semi-specific in nature. The stress regulons are activated as a protection/repair system in response to agents that may provoke genetic or metabolic damage to the cells.

A variety of whole-cell sensing systems have been designed and developed over a time span of almost 20 years [9–13]. Microbial biosensors have been extensively used to detect analytes of interest, such as, metals, anions, sugars, drugs, organic pollutants, and quorum sensing signal molecules [13–16]. Examples of metals include mercury [17, 18], chromium [19], lead [20, 21], copper [22], aluminum

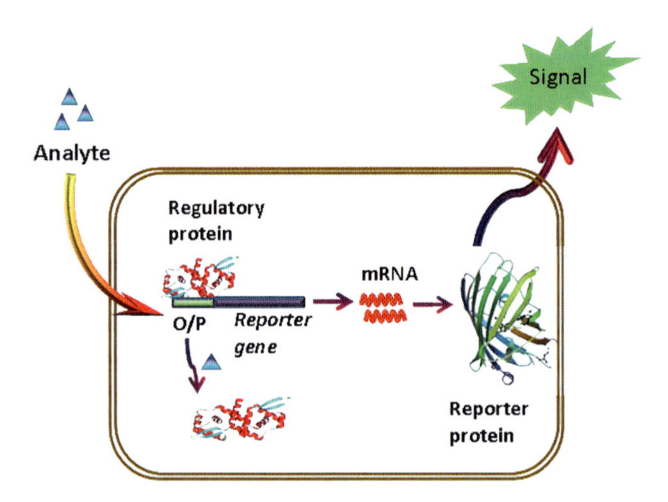

Fig. 2 Schematic representation of an inducible whole-cell sensing system in which the reporter protein is placed under the transcriptional control of a promoter inducible by a target compound. In the absence of analytes, the regulatory protein is bound to the promoter and inhibits the reporter gene transcription. In the presence of analytes, the regulatory protein binds the analyte, and releases itself from the promoter, thus activating the reporter gene transcription and leading to signal production

[23], iron [24, 25], zinc [26], and silver [27]. The detection limits for some of the metals have been reported to be at nanomolar levels. Anions, such as, arsenite, arsenate, and antimonite [28, 29], nitrate [30], and phosphates [31] have been detected by whole-cell sensing. Among organic compounds, cell-based biosensors have been described for alkanes [32], benzene and its derivatives [33, 34], aromatic compounds [35], polychlorinated biphenyls [36], chlorocatechols [37, 38], dihydroxylated(chloro)-biphenyls [39], hydroxylated polychlorinated biphenyls [40], L-arabinose [41], tetracyclines [42], and *N*-acylhomoserine lactones [16]. Furthermore, whole-cell sensing systems have been developed, which are responsive to different kinds of stress, including, heat shock, oxidative stress [43, 45], protein damage [46–48], DNA damage [45, 49–51], and membrane damage [52].

A remarkable feature of living cells is their ability to provide information on the bioavailability and activity of compounds present within the cell. This is particularly important when those are analytes that have been targeted for detection. As compared to isolated sensing proteins, bacterial sensing cells present additional advantages; these include their ability to self-replicate, the lack of need for extensive purification processes before their use, tolerance to relatively harsh environments, and higher stability. Consequently, whole-cell sensing systems are robust and convenient to use. On the other hand, their responses are slower than those of protein-based biosensors and the need for cell culture before use further increases the overall assay time. Additionally, it should be pointed out that self-replication, while advantageous in terms of sensor availability, may be a disadvantage from the analytical point of view because it makes it harder to control the concentration of sensing reagent during the assay. Inducible whole-cell sensing systems are also very attractive compared to chromatographic and spectroscopic techniques and bioassays in environmental and biomedical sensing. The main reason for that lies in the possibility of analyzing samples directly or upon minimal preparation, thus avoiding the lengthy and costly sample pretreatment steps needed with the above methods. Additionally, whole-cell sensing does not require expensive instrumentation and highly-trained technical personnel. A further advantage of whole-cell biosensors is that they are amenable to miniaturization and automation, thus enabling multiplex analysis, high-throughput screening, and field applications.

Despite their advantageous analytical features, whole-cell biosensors have been restricted to use in laboratory settings due to limitations posed by the need for keeping the sensing cells alive and providing them with an environment rich in nutrients, oxygen, etc. One of the major concerns that are still unresolved is the "shelf-life" of bacterial sensor strains. The effective storage of bacterial biosensing systems requires the preservation of the viability as well as activity and analytical performance of the cells. The ability to keep reporter bacteria at ambient temperature for long periods of time without the need for any special requirements, while maintaining their analytical characteristics, is still a major challenge. Preservation, long-term storage, and portability of sensing bacteria are some of the critical features that need to be achieved for on-site applications. To that end, several approaches are currently being employed, which include drying, continuous cultivation, immobilization, and sporulation methods. The features, advantages, and disadvantages, as well as new advances of these methods, as related to the stability and potential on-site use of luminescent whole-cell sensing systems, are critically discussed in this chapter.

2 Drying Methods

Freeze-drying is one of the most common methods for storing whole-cell sensing bacteria and other microbial cultures. Although various methods are used for freeze-drying microorganisms, there are some fundamental common steps. Specifically, cells are grown to an optimal growth phase before they can be mixed with drying preservation agents. Next, the mixture is frozen at low pressure to allow for sublimation of frozen water. These desiccated cells can then be stored for a period of time, and rehydrated when required for further sensing use. However, the survival rate of freeze-dried cells is very low and decreases during long-term storage [53]. Therefore, to ensure revival of a sufficient amount of cells, a relatively high amount of cells is required for the freeze-drying process. Optimal conditions for survival and stability of freeze-dried cells vary depending upon the organism. In general, the most critical parameters are growth phase before drying, growth media, desiccation tolerance, preservation agents, and rehydration methods.

The growth phase ensuring the highest recovery after freeze-drying may vary. For example, it was observed that *Lactobacillus rhamnosus* stationary phase cells had the highest recovery of 31–50%, while only 2% of the early log phase cells of the same species could survive the drying process [54]. In contrast, the survivability of *Sinorhizobium* and *Bradyrhizobium* cells was reported to be highest at the lag phase of growth [55]. The optimal cell concentration before drying is also known to affect the survival rate [56]. As far as the growth media composition is concerned, sucrose has been shown to enhance the viability as well as to maintain the activity and performance of freeze-dried sensing cells. For example, the activity of freshly cultured whole-cell biosensing systems for phenolic compounds was compared to that of the same cells after they were freeze-dried in the presence of 10% glucose, 12% sucrose, or 10% glucose + 12% sucrose, and stored at −20 and −70°C for several months. It was observed that cells freeze-dried in the presence of sucrose and stored at −70°C exhibited the highest relative activity when challenged with phenolic compounds [57]. In another study, Carvalho et al. showed that the viability of freeze-dried cells was higher when the cells were grown in the presence of mannose, as compared to fructose, lactose, or glucose [58, 59]. A study by Streeter et al. indicated that survival of *Bradyrhizobium japonicum* could be significantly enhanced by the presence of trehalose in the growth medium [60]. However, Cho et al. reported that the luminescence activity of Janthinobacterium lividum increased due to the addition of trehalose to the growth medium just prior to freeze-drying [61].

It is important to note that survivability does not necessarily correlate with the activity and analytical performance of the cells. As an example, high salinity of the growth medium is known to enhance the viability of the freeze-dried biosensing bacterial cells, while not necessarily ensuring maintenance of their activity. Pedazhur et al. showed that the cell activity not only depends on the salinity but also the type of growth medium used [62]. Preservation additives are usually mixed along with the cells prior to the freeze-drying process. These include glycol, glycerol, and sulfoxide derivatives such as dimethyl sulfoxide. These compounds are known to reduce the concentration of salt around the cells and induce partial dehydration during the freezing process. Some cryoprotective agents adsorb on the microbial surface and cause the removal of water from the cell [63]. Additives that form a saturated liquid with very high viscosity upon freezing are known as amorphous glass forming protective agents. Ice and salt crystals that form during the freezing process can damage the cell wall of the microorganism. Such highly viscous compounds reduce mobility, keep the structure of ice amorphous within and around the cell, and prevent the harmful waste products produced by the bacteria from concentrating around the cell [64–66].

The stability of freeze-dried bioluminescent bacteria under ambient temperature conditions was studied by Ulitzur et al. [8]. The freeze-dried sensing cells were stored at 25 and 30°C for 3, 5, and 7 days, reconstituted after these times, and evaluated for their sensing ability to their target analytes, namely cadmium and parathion. The results obtained were compared to those shown by sensing cells from the same batch that had been kept at 4°C. It was concluded that storage for up to 7 days, both, at 25 and 30°C, did not lead to a significant reduction in overall luminescence, rather the sensing bacteria maintained their initial analytical performance [8]. In another study by Stocker et al., arsenite/arsenate sensing cells were applied to a paper strip, vacuum dried, and stored at 20, 4, or 30°C [67]. The data collected showed that the performance of arsenite/arsenate sensing bacterial cells was maintained for at least 2 months when stored in the tested conditions. Notably, freeze-drying enables easy and inexpensive shipping of sensing bacteria at room temperature, thus reducing the high operational costs posed by shipping in a chilled container. On the other hand, cost and complexity of the technique represent some of the disadvantages of this preservation method.

3 Immobilization Methods

Immobilization of cells is extensively utilized for storage and preservation of whole-cell biosensing systems. Various immobilization techniques have been developed, including encapsulation in polymer gels, entrapment in different types of matrices, and adsorption, as well as covalent attachment to solid supports, such as fiber optics, microchips, microtiter plates, membranes, and glass slides (Fig. 3). In that regard, the immobilization of cells within polymeric matrices either of organic or inorganic nature has received much attention and has been the focus of numerous studies. In order to keep optimum biostability and efficiency of whole-cell sensing systems, the matrix used for immobilization should keep the cells

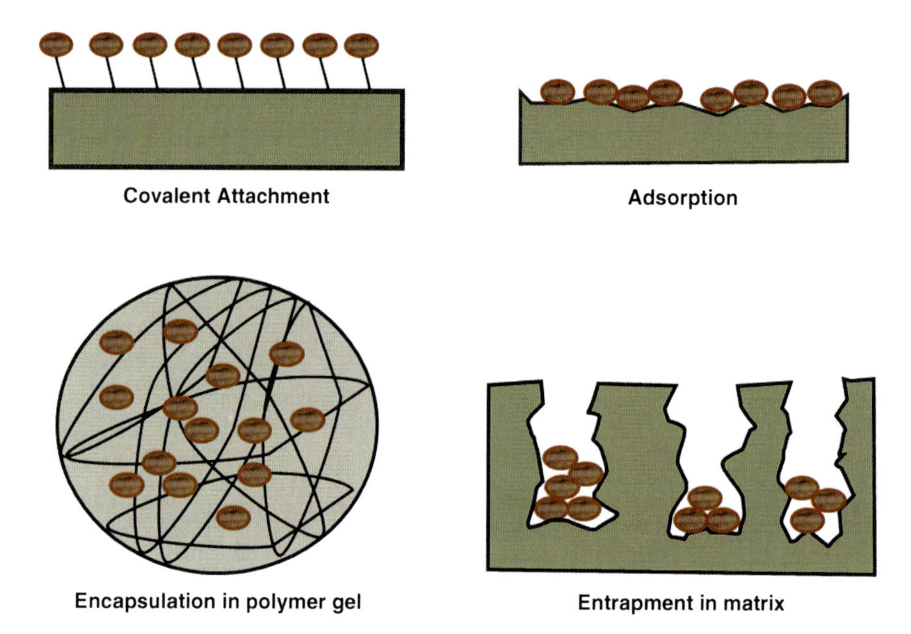

Fig. 3 Schematic representing various immobilization methods of whole-cell sensing systems

isolated from each other. This prevents self aggregation, and protects the cells from microbial attack, while providing essentially the same local aqueous microenvironment of biological media. For a matrix encapsulation/entrapment method to be effective, certain requirements need to be met. The biosensing cells' response may be affected by the environmental stress that the cells are subjected to. Therefore, the conditions associated with preparation of the matrix and entrapment of the cells should be sufficiently mild to keep the cells alive and active in such a manner that the reproducibility of the method is not compromised. Ideally, prolonged storage should result in no inactivation and no growth of the encapsulated cells during that time. No growth is essential to maintain a constant number of cells, and thus the reproducibility of the sensing system. This is especially challenging since *Escherichia coli* and most other microorganisms are known to proliferate rapidly. Additionally, minimal background signal and high sensitivity of the sensing cells are parameters that need to be tightly controlled.

3.1 Sol–Gel Entrapment

Sol–gel entrapment offers several advantages regarding the preservation and storage of whole-cell biosensors. The sol–gel matrix is highly porous and possesses mechanical strength as well as chemical and thermal stability that protects the immobilized cells for relatively long periods of time without affecting their analytical performance. Moreover, since polymerization of the sol–gel matrix can be

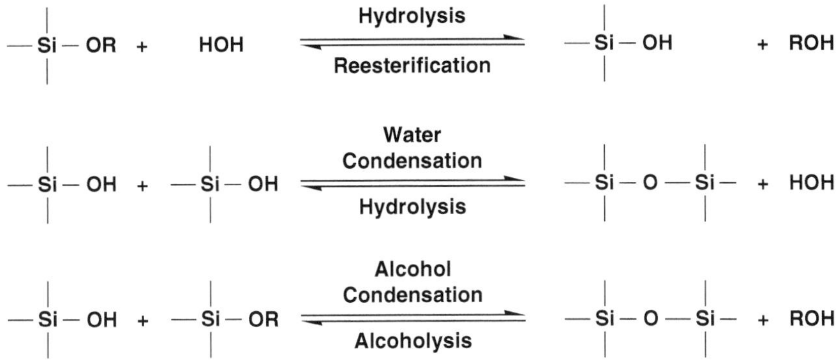

$$\equiv Si{-}OR + HOH \underset{\text{Reesterification}}{\overset{\text{Hydrolysis}}{\rightleftharpoons}} \equiv Si{-}OH + ROH$$

$$\equiv Si{-}OH + \equiv Si{-}OH \underset{\text{Hydrolysis}}{\overset{\text{Water Condensation}}{\rightleftharpoons}} \equiv Si{-}O{-}Si\equiv + HOH$$

$$\equiv Si{-}OH + \equiv Si{-}OR \underset{\text{Alcoholysis}}{\overset{\text{Alcohol Condensation}}{\rightleftharpoons}} \equiv Si{-}O{-}Si\equiv + ROH$$

Fig. 4 Chemical reactions involved in the formation of a sol–gel matrix

carried out at room temperature, the whole cells can be incorporated within the matrix without any special precautions. Silicates are usually employed as precursors in the entrapment of bacteria within sol–gel (Fig. 4). An advantage of silicate-based materials is that they are optically transparent, and thus render themselves as biocompatible matrices for encapsulation of biological optical biosensing systems. Recent studies have showed the successful use of silicates for encapsulation of whole cells in sol–gel. Yu et al. demonstrated that genetically engineered cells capable of detection of organophosphates maintained 95% activity for 2 months after sol–gel encapsulation in a silica matrix [68]. Different entrapment methods using tetramethyl orthosilicate (TMOS) and sodium silicate were also evaluated for their effect on the activity and viability of the cells. The use of TMOS as the starting material for encapsulation led to generation of methanol, which caused the efficiency of the sensing cells to be lower than that in the sodium silicate matrix. However, the activity of encapsulated cells, in both matrices, was higher than that displayed by free cells stored in buffer solution [68]. The effect of release of methanol from TMOS during its polycondensation process on heat shock sensing bacteria was investigated by Premkumar et al. [69, 70]. These researches investigated the response of biosensing bacteria that were encapsulated in TMOS matrix and stored at 4°C for 4 weeks. For that, a strain containing a *lux*-fusion of the promoter of the heat shock gene *grpE* was tested to assess the methanol effect. The luminescence signal from the strain showed a relatively small induction of the heat shock system compared to the deliberate addition of a heat shock inducer, thus suggesting limited effect of methanol on the cells during gel formation. The bacteria behaved as if they were exposed to 0.01 M methanol, which is far below the lethal dose for bacteria [69]. As reported above, one of the concerns in sol–gel encapsulation of whole cells using alkoxides is the release of alcohol, which is toxic to the cells. To address this problem, Amoura et al. employed a strategy based on the formation of alumina gel from preformed boehmite colloids containing glycerol. They demonstrated that the cells were able to maintain their viability for 30 days [71]. It is important to note that sol–gel encapsulated cells need to be kept in wet/humid conditions since drying of the matrix can significantly alter their viability and activity [70]. To circumvent the need for wet/humid conditions during the encapsulation of cells in sol–gel matrices, Tessema et al. proposed a freeze-drying compatible sol–gel process [72]. Freeze-drying is a harsh process in which dehydration may induce a stress response from the cells. To minimize this effect and to stabilize the activity of the cells, protective additives such as trehalose and glucose were added. The authors observed that sol–gel encapsulated cell-based-sensors that were prepared with more than 10% trehalose maintained their activity after the freeze-drying process. These studies revealed that the freeze-dried sol–gel immobilized cells were somewhat stable. The luminescence emission of bacteria harboring a luminescent reporter after 6 weeks at −20°C was 70% of the initial value observed prior to cell storage. Long-term storage of sol–gel encapsulated sensing bacteria is limited by the increase in cross-linking of the network over time, which in turn causes the internal solvent to be expelled from the matrix, resulting in a change of the internal polarity and viscosity, as well as a decrease in the average pore size. The latter, renders the entrapped sensing cells inaccessible to the analyte.

3.2 *Agar Immobilization*

Another attractive method of storage of whole-cell biosensors relies on the use of agar as the immobilization matrix. Agar is a dried hydrophilic colloid extracted from the cell wall of certain algae of the class Rhodophyceae. Specifically, it is an agarose-based polysaccharide able to form hydrogels. Agar has certain advantages over other materials, including its easy preparation, low cost, and good mechanical and acid stability. These characteristics make agar an attractive matrix for the encapsulation of whole-cell biosensors. In that regard, Park et al. evaluated the effect that storing at 4°C during 6 weeks had on agar immobilized *Salmonella typhimurium* cells harboring a plasmid that contained the DNA damage-inducible SOS promoter fused to the promoterless *luxCDABFE* operon from *Photobacterium leiognathi* [73]. It was observed that the sensing bacteria could be stored for up to 4 weeks without any significant loss in sensitivity. In a different study Mitchell and Gu investigated the immobilization of 12 bacterial strains within 2 different matrices, agar and TMOS-based sol–gel in a 96-well plate [74]. For each system, a series of parameters were needed to optimize the responses of the sensing strains employed, i.e., percent agar, matrix drying time, and cell-matrix volume were evaluated. The optimum conditions proved to be 1.5% w/v agar concentration and 100 μL volume for agar immobilization, and 50 μL volume and 20 min drying time for immobilization in the sol–gel matrix. The responses of the cells immediately after encapsulation were similar, regardless of the type of matrix. However, when the reproducibility of the response was tested over a period of 4 weeks, the cells that were immobilized within the agar showed less week-to-week deviation than the cells encapsulated in the sol–gel matrix. This improved reproducibility was possibly due to the presence of nutrients and LB media in the agar matrix.

3.3 *Cryogel Immobilization*

A relatively new method utilized in the immobilization of whole cells involves the use of cryogels. Cryogels are gel matrices that are formed in moderately frozen solutions of monomeric or polymeric precursors. Cryogels typically have interconnected macropores that allow for the diffusion of solutes of any size, as well as mass transport of nano- and even microparticles. The unique structure of cryogels provides them with excellent osmotic, chemical, and mechanical stability, thus making them promising matrices for immobilization of whole cells. Polyvinyl alcohol (PVA) is an example of polymer employed in the preparation of cryogels for cell entrapment. PVA is a biologically compatible, nontoxic, and readily available low-cost polymer. Moreover, this matrix is a nonbrittle gel material and exhibits little abrasive erosion. Lopez-Fouz et al. studied the effect of storage of *Rodococcus fascians* cells in PVA-polyethylene glycol (PEG) cyrogel on the efficacy of hydrolysis of the phytochemical limonin [75]. The encapsulated cells were viable and maintained reproducible response for at least 1 month without nutrient supplementation, when stored at 4°C.

4 Sporulation Methods

Although the previously described methods offer good properties for preservation and storage of whole-cell sensing systems, they still fall short when the bacterial sensors need to be stored for a long period of time and transported to locations where certain facilities, such as refrigeration, are not available, and to environments with harsh conditions, such as high salt, extreme pHs, cold, dry heat, wet heat, drought, etc. Therefore, there is a need for storage and transport methods for whole-cell sensing systems, which are inexpensive, effective in preserving cells' viability and activity in a wide range of conditions, and easy to use. One promising method of preservation of the sensing bacteria involves the formation of spores.

Some bacteria, such as *Bacillus* and *Clostridium,* can adapt to changing environments and harsh conditions by forming highly resistant spores (Fig. 5). The process of spore formation, called sporulation, involves different stages, including initiation, chromosome segregation, sporulation-specific cell division, differential gene expression, and specific signal transduction mechanisms. The result of this process is a dry, dormant, and hardened vehicle designed to preserve the DNA even in very unfavorable conditions. A spore is composed of a set of protective structures arranged in series of concentric shells, whose most important function is to lock the bacterial DNA into a stable crystalline state, excluding any toxic molecules that may be present in the surrounding environment. Resistance of spores depends on three substructures: the core, cortex, and coat of the spore (Fig. 6). The interior compartment contains the DNA, which is complexed with small acid soluble proteins [76]. The core is surrounded by the cortex, which is composed of a layer

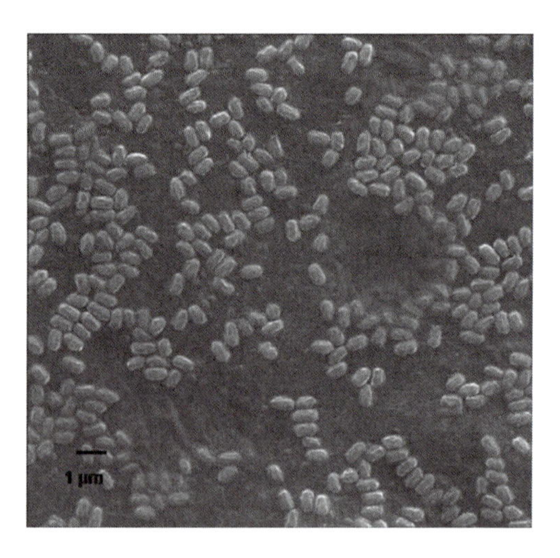

Fig. 5 Scanning electron microscopy (SEM) photograph of *Bacillus subtilis* spores. Magnification 4,500×.*Scale bar*: 1 µm

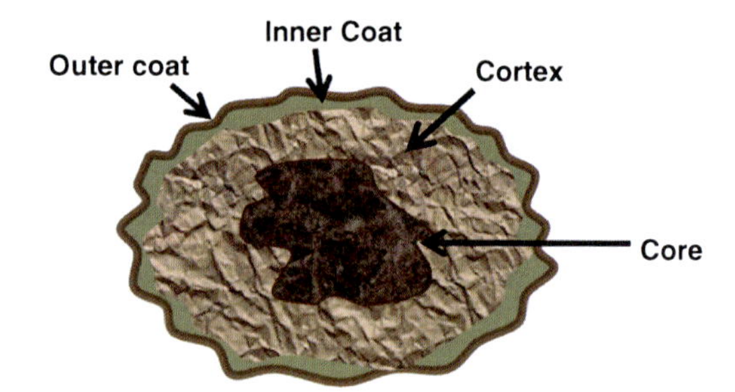

Fig. 6 Schematic of the cross-section of a bacterial spore showing a complex structure comprised of concentric layers. Bacterial DNA is locked in the core

of loosely cross-linked peptidoglycans [77]. Finally, the outer surface, i.e., the coat, is made of multilayered protein shells [78]. Spores are metabolically inactive, highly resilient, and stable for long periods of time. The mature spore released from the mother cell can survive in a metabolically dormant state for hundreds, if not thousands, of years without losing its viability [79]. During the dormant stage, the spores carry out no detectable synthesis or oxidative metabolism, but can acquire normal cell functions within minutes in response to specific germinants in the environment. In spite of their inert state, they can sense even small amounts of nutrients, and respond by germinating to vegetative growing cells [79].

These unique features of spores prompted us to employ them for stabilization of whole-cell biosensors. Specifically, our approach is based on the use of spore-forming bacteria for the development of sensing systems, and the generation of spores as a simple, inexpensive, stable, and resistant way of preservation, storage, and transport of whole-cell biosensors. As mentioned above, spores can be germinated to generate viable and metabolically active cells. In our work, we have used spore-forming bacteria such as *Bacillus subtilis* and *Bacillus megaterium* for development of luminescent sensing systems for two model analytes, namely arsenic and zinc [26]. The two sensing systems were analytically characterized by

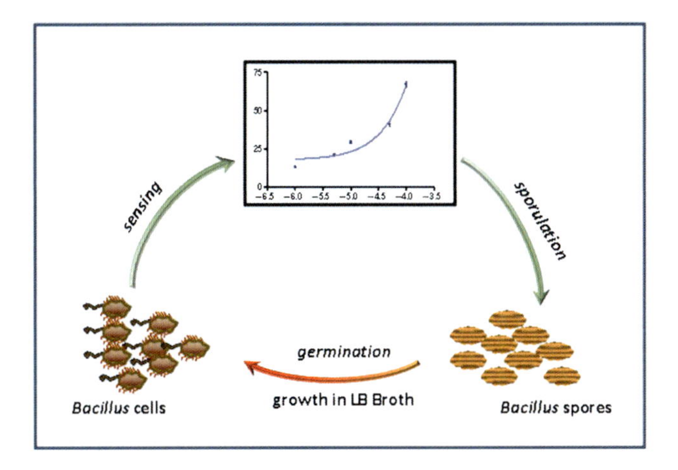

Fig. 7 Sensing–sporulation–germination cycles employing spore-based whole-cell sensing systems

exposing them to varying concentrations of target analytes. Subsequently, these sensing cells were converted to spores and stored for a period of time at room temperature. The spores were then germinated to vegetative cells, which were employed for sensing. We demonstrated that these spore-based sensing systems retained their analytical performance in terms of detection limits, dynamic range, and reproducibility after storage at room temperature for up to 24 months, as well as after three cycles where the cells alternated between being dormant or active, i.e., sporulation–germination cycles. Figure 7 shows a schematic representation of the experimental setup including sensing–sporulation–germination cycles. The potential advantages of using spore-based biosensors include heat, cold, and drought stability, thus permitting their application in extreme environmental conditions, safety record established, for example, through use of spores as probiotics [80] and simple and economic production of spores based on commonly used microbiological protocols.

An interesting study that employed spores for real-time measurement of bacterial contamination of platelets concentrates was performed by Rotman et al. [81]. The target bacteria, when present, induced the formation of a spore germinant by hydrolyzing a germinogenic substrate. This germinant revived the spores, and produced acetyl esterase, which was detected by employing a fluorescent substrate. Although this system was not used as a storage method for cell-based sensors, it certainly illustrated the advantages offered by spores in terms of stability and safety. Among other examples demonstrating the advantages of spores is the report by Marston et al. on the effects of long-term storage of spores immobilized in a PVA cryogel on the viability and biosynthesis activity of the vegetative cells [82]. The authors demonstrated that the immobilized spores maintained their viability for at least 1.5 years. Another application was illustrated by Lee et al., when using spores as a sensor for germinants in spatially confined nonbiofouling microwells.

The biospecific interactions between biotinylated *B. subtilis* spores and streptavidin were used for selective attachment of spores to the bottom of the microwells [83]. The viability of the patterned spores was confirmed by inducing germination. The authors showed that, in principle, the spores could be spatially confined and stored in microwells and could be revived in the presence of germinating agents, thus enabling detection of these compounds. Surface display of proteins on spore surfaces is also one of the emerging techniques that may be employed in the development of sturdy bacterial biosensors. As an example, Du et al. were able to display a specific antibody and the GFP on the surface of *Bacillus thuringiensis* [84]. The spore surface display system may provide a durable supporting matrix similar to that of chemical polymer beads, which is easy and economical to produce in large quantities.

5 Conclusions

Fluorescent and bioluminescent whole-cell sensing systems proved to be valuable analytical tools, which allow sensitive, rapid, and inexpensive quantitative detection of target analytes. Depending on the required applications, whole-cell-based biosensors can be designed to be nonspecific, semispecific, or highly specific/selective. They are amenable to miniaturization and, therefore, they would be ideal for on-site analysis. However, a main limitation for that is the relatively short shelf-life of living sensing cells. A good amount of work has been performed towards the development of efficient and reliable methods for preservation and storage of whole-cell sensing systems. Freeze-drying of the cells and various immobilization techniques to entrap the cells have been employed to maintain the viability and activity of the sensing systems. Each of these methods offers some unique advantages that make it attractive for cell preservation. Nevertheless, each of them has some drawbacks. For instance, although freeze-drying provides an easy and reliable method of preservation, it is an expensive and time-consuming process. Immobilization within polymeric matrices generally provides good physical protection of the cells and mechanical stability. However, biodegradation, bacterial growth, and performance problems may arise, depending on the nature of the polymer used. Additionally, the preservation time afforded by these methods is relatively short, up to a few months, and control of storage temperature appears to be needed or preferred. On the other hand, construction of spore-based whole-cell sensing system is a new method that has been effectively employed for long-term (up to 2 years) preservation of sensing cells at room temperature, as well as at 37°C, 28°C with relative humidity >80%, −20°C, and desiccated. Spores are also attractive as vehicles for transport of whole-cell sensing systems out of the laboratory and are amenable for incorporation into portable devices. To that end, spore-based whole-cell sensing systems could be integrated into miniaturized analytical platforms, such as microcentrifuge microfluidics platforms, microchips, and optic fibers. It is envisaged that such spore-based sensing systems can be packaged along with all needed reagents, thus generating micrototal analytical systems (μ-TAS), which would provide a viable tool for environmental and clinical on-site analysis. Moreover, the natural hardiness and ruggedness of spores are expected to enable the use of whole-cell sensing systems in harsh environments. This would be particularly beneficial where appropriate storage and transport facilities are insufficient or unavailable.

Acknowledgments This work was partly supported by the National Science Foundation, Grants CHE-0416553 and CHE-0718844, the National Institute of Environmental Health Sciences, Grant P42ES07380, and the United States–Israel Binational Agricultural Research and Development Fund, Grant US-3864-06. In addition, SD would like to acknowledge support from the Gill Endowment of the College of Arts & Sciences for a Gill Professorship.

References

1. Deo SK, Daunert S (2001) Luminescent proteins from *Aequorea victoria*: applications in drug discovery and in high throughput analysis. Fresenius' J Anal Chem 369(3):258–266
2. Meighen EA (1991) Molecular biology of bacterial bioluminescence. Microbiol Mol Biol Rev 55(1):123–142
3. Aboul-Enein HY, Stefan R-I, van Staden JF (1999) Chemiluminescence-based (bio)sensors – an overview. Crit Rev Anal Chem 29:323–331
4. Bulich AA, Isenberg DL (1981) Use of the luminescent bacterial system for the rapid assessment of aquatic toxicity. ISA Trans 20(1):29–33
5. Dalzell DJB, Alte S, Aspichueta E, de la Sota A, Etxebarria J, Gutierrez M, Hoffmann CC, Sales D, Obst U, Christofi N (2002) A comparison of five rapid direct toxicity assessment methods to determine toxicity of pollutants to activated sludge. Chemosphere 47(5):535–545
6. Preston S, Coad N, Townend J, Killham K, Paton GI (2000) Biosensing the acute toxicity of metal interactions: are they additive, synergistic, or antagonistic? Environ Toxicol Chem 19(3):775–780
7. Leveau JHJ, Lindow SE (2002) Bioreporters in microbial ecology. Curr Opin Microbiol 5 (3):259–265
8. Ulitzur S, Lahav T, Ulitzur N (2002) A novel and sensitive test for rapid determination of water toxicity. Environ Toxicol 17(3):291–296
9. Gooding JJ (2003) Biosensors for detecting metal ions: new trends. Aust J Chem 56(2–3):159
10. Lei Y, Chen W, Mulchandani A (2006) Microbial biosensors. Anal Chim Acta 568(1–2): 200–210
11. Verma N, Singh M (2005) Biosensors for heavy metals. Biometals 18(2):121–129
12. Wanekaya AK, Chen W, Mulchandani A (2008) Recent biosensing developments in environmental security. J Environ Monitor 10(6):703–712
13. Yagi K (2007) Applications of whole-cell bacterial sensors in biotechnology and environmental science. Appl Microbiol Biotechnol 73:1251–1258
14. Feliciano J, Pasini P, Deo SK, Daunert S (2006) Photoproteins as reporters in whole-cell sensing. In: Daunert S, Deo SK (eds) Photoproteins in bioanalysis. Wiley-VCH, Weinheim, pp 131–154
15. Galluzzi L, Karp M (2006) Whole cell strategies based on lux genes for high throughput applications toward new antimicrobials. Comb Chem High Throughput Screen 9:501–514
16. Kumari A, Pasini P, Deo SK, Flomenhoft D, Shashidhar H, Daunert S (2006) Biosensing systems for the detection of bacterial quorum signaling molecules. Anal Chem 78(22): 7603–7609

17. Ivask A, Hakkila K, Virta M (2001) Detection of organomercurials with sensor bacteria. Anal Chem 73(21):5168–5171
18. Virta M, Lampinen J, Karp M (1995) A luminescence-based mercury biosensor. Anal Chem 67(3):667–669
19. Peitzsch N, Eberz G, Nies DH (1998) Alcaligenes eutrophus as a bacterial chromate sensor. Appl Environ Microbiol 64(2):453–458
20. Shetty RS, Deo SK, Shah P, Sun Y, Rosen BP, Daunert S (2003) Luminescence-based whole-cell-sensing systems for cadmium and lead using genetically engineered bacteria. Anal Bioanal Chem 376(1):11–17
21. Tauriainen S, Karp M, Chang W, Virta M (1998) Luminescent bacterial sensor for cadmium and lead. Biosens Bioelectron 13(9):931–938
22. Shetty RS, Deo SK, Liu Y, Daunert S (2004) Fluorescence-based sensing system for copper using genetically engineered living yeast cells. Biotechnol Bioeng 88(5):664–670
23. Guzzo J, Guzzo A, Dubow MS (1992) Characterization of the effects of aluminum on luciferase biosensors for the detection of ecotoxicity. Toxicol Lett 64–65:687–693
24. Boyanapalli R, Bullerjahn GS, Pohl C, Croot PL, Boyd PW, McKay RML (2007) Luminescent whole-cell cyanobacterial bioreporter for measuring Fe availability in diverse marine environments. Appl Environ Microbiol 73(3):1019–1024
25. Joyner DC, Lindow SE (2000) Heterogeneity of iron bioavailability on plants assessed with a whole-cell GFP-based bacterial biosensor. Microbiology 146(10):2435–2445
26. Date A, Pasini P, Daunert S (2007) Construction of spores for portable bacterial whole-cell biosensing systems. Anal Chem 79(24):9391–9397
27. Riether K, Dollard MA, Billard P (2001) Assessment of heavy metal bioavailability using Escherichia coli zntAp::lux and copAp::lux-based biosensors. Appl Microbiol Biotechnol 57(5):712–716
28. Ramanathan S, Shi W, Rosen BP, Daunert S (1997) Sensing antimonite and arsenite at the subattomole level with genetically engineered bioluminescent bacteria. Anal Chem 69(16):3380–3384
29. Rothert A, Deo SK, Millner L, Puckett LG, Madou MJ, Daunert S (2005) Whole-cell-reporter-gene-based biosensing systems on a compact disk microfluidics platform. Anal Biochem 342 (1):11–19
30. Taylor CJ, Bain LA, Richardson DJ, Spiro S, Russell DA (2004) Construction of a whole-cell gene reporter for the fluorescent bioassay of nitrate. Anal Biochem 328(1):60–66
31. Dollard M-A, Billard P (2003) Whole-cell bacterial sensors for the monitoring of phosphate bioavailability. J Microbiol Meth 55(1):221–229
32. Sticher P, Jaspers M, Stemmler K, Harms H, Zehnder A, van der Meer J (1997) Development and characterization of a whole-cell bioluminescent sensor for bioavailable middle-chain alkanes in contaminated groundwater samples. Appl Environ Microbiol 63(10):4053–4060
33. Ikariyama Y, Nishiguchi S, Koyama T, Kobatake E, Aizawa M, Tsuda M, Nakazawa T (1997) Fiber-optic-based biomonitoring of benzene derivatives by recombinant E. coli bearing luciferase gene-fused TOL-plasmid immobilized on the fiber-optic end. Anal Chem 69 (13):2600–2605
34. Stiner L, Halverson LJ (2002) Development and characterization of a green fluorescent protein-based bacterial biosensor for bioavailable toluene and related compounds. Appl Environ Microbiol 68(4):1962–1971
35. Phoenix A, Keane A, Patel H, Bergeron S, Ghoshal PC, Lau K (2003) Characterization of a new solvent-responsive gene locus in *Pseudomonas putida* F1 and its functionalization as a versatile biosensor. Environ Microbiol 5(12):1309–1327
36. Layton AC, Muccini M, Ghosh MM, Sayler GS (1998) Construction of a bioluminescent reporter strain to detect polychlorinated biphenyls. Appl Environ Microbiol 64(12): 5023–5026
37. Guan X, d'Angelo E, Luo W, Daunert S (2002) Whole-cell biosensing of 3-chlorocatechol in liquids and soils. Anal Bioanal Chem 374(5):841–847

38. Guan X, Ramanathan S, Garris JP, Shetty RS, Ensor M, Bachas LG, Daunert S (2000) Chlorocatechol detection based on a clc operon/reporter gene system. Anal Chem 72(11):2423–2427
39. Feliciano J, Xu S, Guan X, Lehmler H-J, Bachas L, Daunert S (2006) ClcR-based biosensing system in the detection of cis-dihydroxylated (chloro-)biphenyls. Anal Bioanal Chem 385 (5):807–813
40. Turner K, Xu S, Pasini P, Deo S, Bachas L, Daunert S (2007) Hydroxylated polychlorinated biphenyl detection based on a genetically engineered bioluminescent whole-cell sensing system. Anal Chem 79(15):5740–5745
41. Shetty RS, Ramanathan S, Badr IHA, Wolford JL, Daunert S (1999) Green fluorescent protein in the design of a living biosensing system for l-arabinose. Anal Chem 71(4):763–768
42. Bahl MI, Hansen LH, Licht TR, Sorensen SJ (2004) In vivo detection and quantification of tetracycline by use of a whole-cell biosensor in the rat intestine. Antimicrob Agents Chemother 48(4):1112–1117
43. Belkin S, Smulski D, Vollmer A, Van Dyk T, LaRossa R (1996) Oxidative stress detection with Escherichia coli harboring a katG'::lux fusion. Appl Environ Microbiol 62(7):2252–2256
44. Lee H, Gu M (2003) Construction of a sodA::luxCDABE fusion Escherichia coli: comparison with a katG fusion strain through their responses to oxidative stresses. Appl Microbiol Biotechnol 60(5):577–580
45. Mitchell RJ, Gu MB (2004) An Escherichia coli biosensor capable of detecting both genotoxic and oxidative damage. Appl Microbiol Biotechnol 64(1):46–52
46. Cha HJ, Srivastava R, Vakharia VN, Rao G, Bentley WE (1999) Green fluorescent protein as a noninvasive stress probe in resting Escherichia coli cells. Appl Environ Microbiol 65(2):409–414
47. Sagi E, Hever N, Rosen R, Bartolome AJ, Rajan Premkumar J, Ulber R, Lev O, Scheper T, Belkin S (2003) Fluorescence and bioluminescence reporter functions in genetically modified bacterial sensor strains. Sens Actuators B Chem 90(1–3):2–8
48. Van Dyk TK, Majarian WR, Konstantinov KB, Young RM, Dhurjati PS, LaRossa RA (1994) Rapid and sensitive pollutant detection by induction of heat shock gene-bioluminescence gene fusions. Appl Environ Microbiol 60(5):1414–1420
49. Kostrzynska M, Leung KT, Lee H, Trevors JT (2002) Green fluorescent protein-based biosensor for detecting SOS-inducing activity of genotoxic compounds. J Microbiol Meth 48(1):43–51
50. Norman A, Hestbjerg HL, Sorensen SJ (2005) Construction of a ColD cda promoter-based SOS-green fluorescent protein whole-cell biosensor with higher sensitivity toward genotoxic compounds than constructs based on recA, umuDC, or sulA promoters. Appl Environ Microbiol 71(5):2338–2346
51. Vollmer A, Belkin S, Smulski D, Van Dyk T, LaRossa R (1997) Detection of DNA damage by use of Escherichia coli carrying recA'::lux, uvrA'::lux, or alkA'::lux reporter plasmids. Appl Environ Microbiol 63(7):2566–2571
52. Bechor O, Smulski DR, Van Dyk TK, LaRossa RA, Belkin S (2002) Recombinant microorganisms as environmental biosensors: pollutants detection by Escherichia coli bearing fabA'::lux fusions. J Biotechnol 94(1):125–132
53. Bozoglu TF, Özilgen M, Bakir U (1987) Survival kinetics of lactic acid starter cultures during and after freeze drying. Enzyme Microb Technol 9(9):531–537
54. Corcoran RP, Ross GF, Stanton FC (2004) Comparative survival of probiotic lactobacilli spray-dried in the presence of prebiotic substances. J Appl Microbiol 96(5):1024–1039
55. Boumahdi M, Mary P, Hornez JP (1999) Influence of growth phases and desiccation on the degrees of unsaturation of fatty acids and the survival rates of rhizobia. J Appl Microbiol 87(4):611–619
56. Costa E, Usall J, Teixidó N, Garcia N, Viñas I (2000) Effect of protective agents, rehydration media and initial cell concentration on viability of *Pantoea agglomerans* strain CPA-2 subjected to freeze-drying. J Appl Microbiol 89(5):793–800

57. Shin HJ, Park HH, Lim WK (2005) Freeze-dried recombinant bacteria for on-site detection of phenolic compounds by color change. J Biotechnol 119(1):36–43
58. Carvalho AS, Silva J, Ho P, Teixeira FP, Malcata X, Gibbs P (2004) Effects of various sugars added to growth and drying media upon thermotolerance and survival throughout storage of freeze-dried *Lactobacillus delbrueckii* ssp. *bulgaricus*. Biotechnol Prog 20(1):248–254
59. Carvalho AS, Silva J, Ho P, Teixeira P, Malcata FX, Gibbs P (2003) Effect of various growth media upon survival during storage of freeze-dried *Enterococcus faecalis* and *Enterococcus durans*. J Appl Microbiol 94(6):947–952
60. Streeter JG (2003) Effect of trehalose on survival of *Bradyrhizobium japonicum* during desiccation. J Appl Microbiol 95(3):484–491
61. Cho J-C, Park K-J, Ihm H-S, Park J-E, Kim S-Y, Kang I, Lee K-H, Jahng D, Lee D-H, Kim S-J (2004) A novel continuous toxicity test system using a luminously modified freshwater bacterium. Biosens Bioelectron 20(2):338–344
62. Pedahzur R, Rosen R, Belkin S (2004) Stabilization of recombinant bioluminescent bacteria for biosensor applications. Cell Preserv Technol 2(4):260–269
63. Hubálek Z (2003) Protectants used in the cryopreservation of microorganisms. Cryobiology 46(3):205–229
64. Leslie S, Israeli E, Lighthart B, Crowe J, Crowe L (1995) Trehalose and sucrose protect both membranes and proteins in intact bacteria during drying. Appl Environ Microbiol 61 (10):3592–3597
65. Linders LJM, Wolkers WF, Hoekstra FA, van't Riet K (1997) Effect of added carbohydrates on membrane phase behavior and survival of dried Lactobacillus plantarum. Cryobiology 35 (1):31–40
66. Lodato P, Segovia de Huergo M, Buera MP (1999) Viability and thermal stability of a strain of Saccharomyces cerevisiae freeze-dried in different sugar and polymer matrices. Appl Microbiol Biotechnol 52(2):215–220
67. Stocker J, Balluch D, Gsell M, Harms H, Feliciano J, Daunert S, Malik KA, van der Meer JR (2003) Development of a set of simple bacterial biosensors for quantitative and rapid measurements of arsenite and arsenate in potable water. Environ Sci Technol 37(20):4743–4750
68. Yu D, Volponi J, Chhabra S, Brinker CJ, Mulchandani A, Singh AK (2005) Aqueous sol-gel encapsulation of genetically engineered Moraxella spp. cells for the detection of organophosphates. Biosens Bioelectron 20(7):1433–1437
69. Premkumar JR, Lev O, Rosen R, Belkin S (2001) Encapsulation of luminous recombinant *E. coli* in sol-gel silicate films. Adv Mater 13(23):1773–1775
70. Premkumar JR, Rosen R, Belkin S, Lev O (2002) Sol-gel luminescence biosensors: encapsulation of recombinant E. coli reporters in thick silicate films. Anal Chim Acta 462 (1):11–23
71. Amoura M, Nassif N, Roux C, Livage J, Coradin T (2007) Sol-gel encapsulation of cells is not limited to silica: long-term viability of bacteria in alumina matrices. Chem Commun 39:4015–4017
72. Tessema DA, Rosen R, Pedazur R, Belkin S, Gun J, Ekeltchik I, Lev O (2006) Freeze-drying of sol-gel encapsulated recombinant bioluminescent E. coli by using lyo-protectants. Sens Actuators B Chem 113(2):768–773
73. Park KS, Baumstark-Khan C, Rettberg P, Horneck G, Rabbow E, Gu MB (2005) Immobilization as a technical possibility for long-term storage of bacterial biosensors. Radiat Environ Biophys 44(1):69–71
74. Mitchell RJ, Gu MB (2006) Characterization and optimization of two methods in the immobilization of 12 bioluminescent strains. Biosens Bioelectron 22(2):192–199
75. López-Fouz M, Pilar-Izquierdo MC, Martínez-Mayo I, Ortega N, Pérez-Mateos M, Busto MD (2007) Immobilization of Rhodococcus fascians cells in poly(vinyl alcohol) cryogels for the debittering of citrus juices. J Biotechnol 131(2, Suppl 1):S104
76. Setlow P (1995) Mechanisms for the prevention of damage to DNA in spores of Bacillus species. Annu Rev Microbiol 49(1):29–54

77. Popham DL (2002) Specialized peptidoglycan of the bacterial endospore: the inner wall of the lockbox. Cell Mol Life Sci 59(3):426–433
78. Takamatsu H, Watabe K (2002) Assembly and genetics of spore protective structures. Cell Mol Life Sci 59(3):434–444
79. Santo LY, Doi RH (1974) Ultrastructural analysis during germination and outgrowth of bacillus subtilis spores. J Bacteriol 120(1):475–481
80. Hong HA, Duc LH, Cutting SM (2005) The use of bacterial spore formers as probiotics. FEMS Microbiol Rev 29(4):813–835
81. Rotman B, Cote MA (2003) Application of a real-time biosensor to detect bacteria in platelet concentrates. Biochem Biophys Res Commun 300(1):197–200
82. Marston CK, Hoffmaster AR, Wilson KE, Bragg SL, Plikaytis B, Brachman P, Johnson S, Kaufmann AF, Popovic T (2005) Effects of long-term storage on plasmid stability in Bacillus anthracis. Appl Environ Microbiol 71(12):7778–7780
83. Lee K-B, Hwan Jung Y, Lee Z-W, Kim S, Choi IS (2007) Biospecific anchoring and spatially confined germination of bacterial spores in non-biofouling microwells. Biomaterials 28 (36):5594–5600
84. Du C, Chan WC, McKeithan TW, Nickerson KW (2005) Surface display of recombinant proteins on *Bacillus thuringiensis* spores. Appl Environ Microbiol 71(6):3337–3341

Adv Biochem Engin/Biotechnol (2010) 117: 77–84
DOI: 10.1007/10_2009_17

Published online: 20 January 2010

Electrochemical Cell-Based Sensors

Eliora Z. Ron and Judith Rishpon

Abstract One of the recently developed monitoring technologies involves the use of whole cell biosensors. Such biosensors can be constructed to detect expression of genes of interest and the effect of the environment on this expression. These biosensors are essential for monitoring environmental stress, such as general toxicity or specific toxicity caused by pollutants. Currently, a large spectrum of microbial biosensors have been developed that enable the monitoring of gene expression by measuring light, fluorescence, color, or electric current. The electrochemical monitoring is of special interest for in situ measurements as it can be performed using simple, compact, and mobile equipment and is easily adaptable for online measurements. Here we survey the potential application of electrochemical biosensors with special focus on monitoring environmental pollution.

Keywords Monitoring environmental pollution • Electrochemical biosensors

Contents

E.Z. Ron (✉) and J. Rishpon
Department of Molecular Microbiology and Biotechnology, Faculty of Life Sciences, Tel-Aviv University, Tel-Aviv 69978, Israel
e-mail: eliora@post.tau.ac.il

1 General Principles

Electrochemical whole cell sensors measure the activity of reporter genes whose products can be determined electrochemically. As an example – β-galactosidase, the product of the *lacZ* gene, can be quantified electrochemically (as well as colorimetrically) and used for determining the relative concentration of bacteria expressing *lacZ*, or in gene fusion downstream to promoters of interest.

2 Advantages of Electrochemical Cell Sensors

The advantages of electrochemical measurements include:

- Speed
- Can be used online and in situ
- Sensitive – as a rule, electrochemical determinations are highly sensitive and can detect relatively low activities
- Reproducible
- easy to manipulate and multiply
- Ease of miniaturization
- Quantification is independent of the presence of oxygen and can be performed in turbid solutions

Because of these properties, electrochemical biosensors are especially useful for monitoring environmental pollution, as they can be used in the site that has to be monitored and monitor accurately in water, turbid water, or soil.

3 A Typical Electrochemical Sensor Cell and Its Uses

A typical system employs a compact analyzer and disposable electrodes, and enables simultaneous measurements of several samples [1]. The system and one electrochemical cell are shown in Fig. 1. As already noted above, the system can be miniaturized and even monitored by the use of a hand-held computer (Fig. 2) [2].

Electrochemical measurements of whole cell sensor can determine the activity of a native, natural gene. As an example – the level of β-galactosidase activity can reflect the concentration of bacteria that can ferment lactose, and can be employed for uses such as determination of coliforms in water [3, 4]. The same procedure can also be used to determine the water quality, as the expression of β-galactosidase will be lowered by toxic substances in the water.

Another more common use involves genetically manipulated genes. Thus, it is possible to fuse the gene coding for the reporter downstream to a promoter whose activity we want to monitor. As an example – the *lacZ* gene coding for β-galactosidase can be fused downstream to a promoter that responds to the presence of heavy

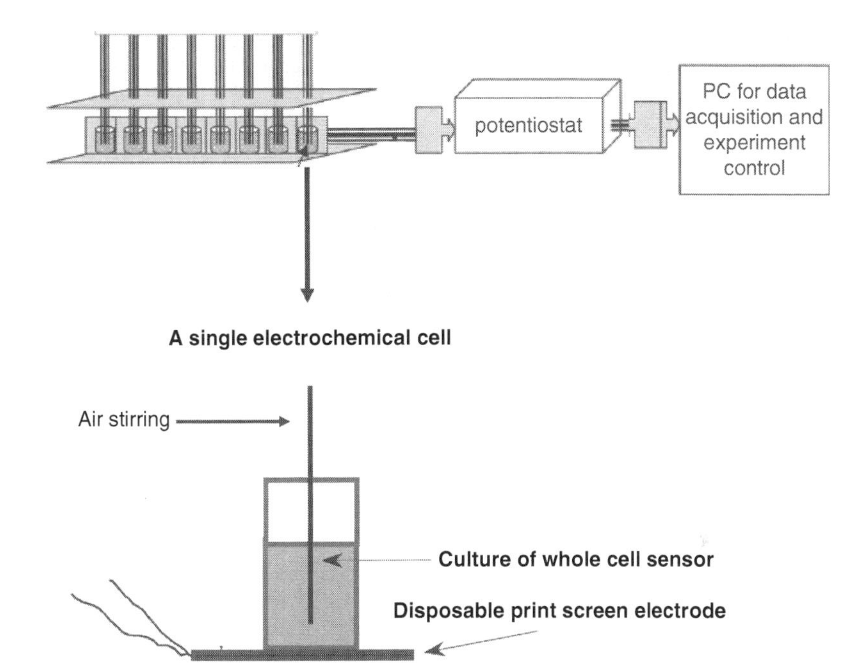

Fig. 1 Schematic presentation of an electrochemical sensing device

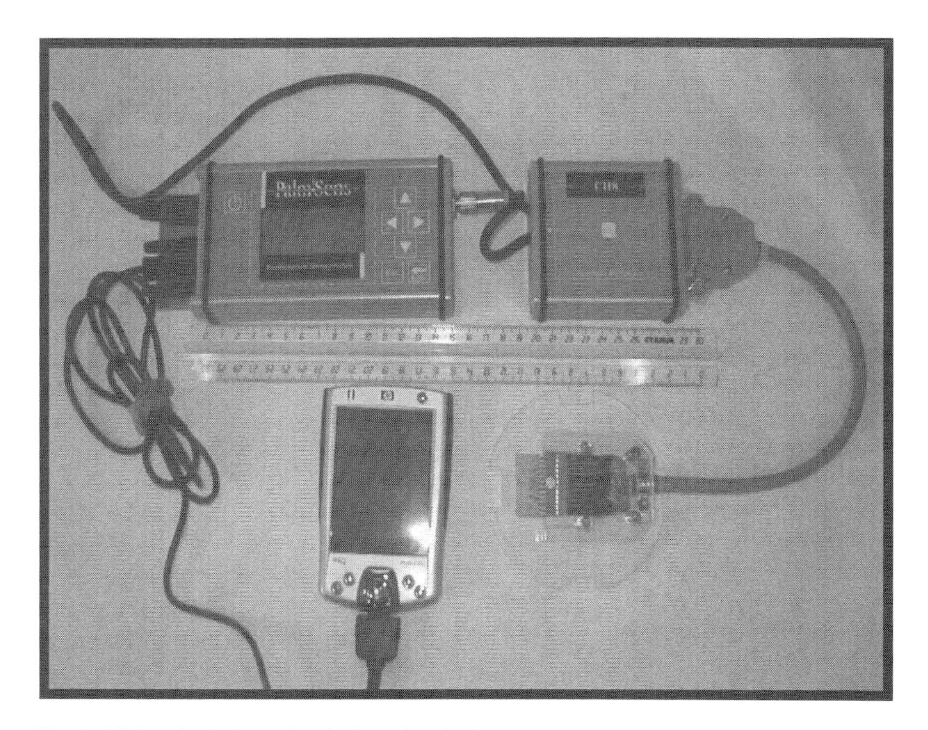

Fig. 2 Miniaturized electrochemical sensing device

metals [5]. This genetic construct can be used for monitoring the concentration of heavy metals. Moreover, it is possible to move such an engineered gene into several types of cells and monitor the expression of genes in bacteria, yeasts and even in mammalian tissue cultures [1, 6, 7].

The construction of an efficient and specific biosensor depends on the choice of promoter as well as reporter gene.

4 Choosing a Promoter

The promoter used is dictated by the environmental factor that needs to be monitored. Thus, promoters are chosen that respond to heavy metals, organic solvents, temperature, etc.

Several factors are considered for choosing a responsive promoter, the two major ones being sensitivity and specificity. As a rule, biological systems are very sensitive and bacterial gene promoters can detect parts per billion of heavy metals and can detect hydrocarbons even as vapor. However, the promoters often respond to groups of compounds rather than to a specific one. For example, promoters that detect cadmium usually also detect all other heavy metals – mercury, zinc, and copper, and some also detect lead [5].

A variety of well characterized promoters is available for genetic manipulations. These include promoters for various heavy metals [6, 8–16], hydrocarbons and organic solvents [7, 17–21], pesticides [11, 22], salicylates [23], various organophosphorus nerve agents [24, 25], mutagens, and genotoxins [26, 27]. There are also available promoters for the evaluation of general toxicity [27–31].

If the required promoter is not available, it is always possible to identify new, suitable promoters. Today, this can be done using transcriptomic or proteomic technologies and identify genes which are under the control of the relevant promoters. For example, promoters responsive to the presence of a certain toxin can be identified by exposure of microbial cultures to the toxin and looking for specific increase in gene expression. The expression of each gene is determined by looking at the whole set up of proteins (proteomes) on two-dimensional gel electrophoresis, or in microarrays which determine the relative transcriptional activity from each gene [32]. Once the responsive genes are identified, their promoters can be amplified by polymerase chain reaction (PCR) and cloned upstream to a suitable reported gene.

5 Choosing a Reporter Gene

The reporter systems are usually enzymes that catalyze an easily monitored reaction. For example, the activity of the enzyme β-galactosidase, which splits β-galactose bonds, can be determined electrochemically. In addition to the widely used β-galactosidase, other enzymes, such as alkaline phosphatase, can also be used

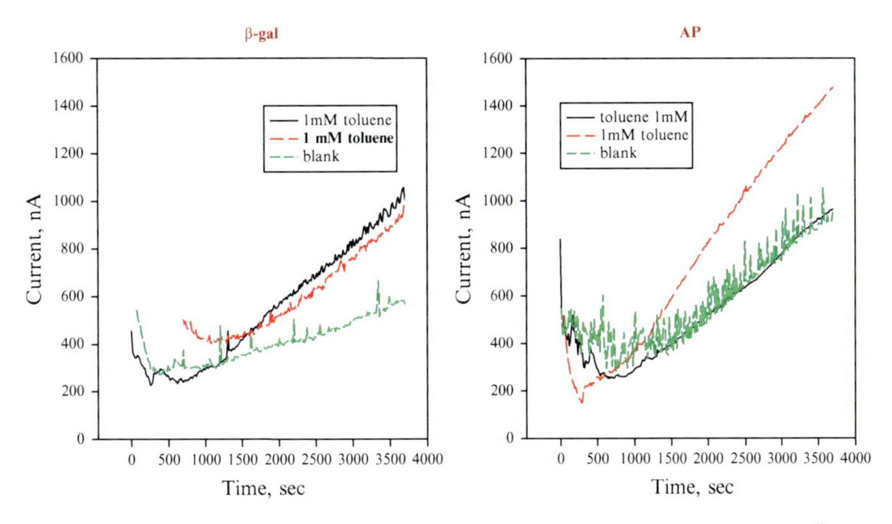

Fig. 3 Monitoring toluene with an electrochemical biosensor with β-galactosidase (*left*) or alkaline phosphatase (*right*) as a reporter gene

[6, 7, 33–35]. The use of several enzymes facilitates a differential sensing of two pollutants by one biosensor.

It should be noted that each reporter has its unique properties that can be useful for monitoring [6, 7]. For example, a comparison between β-galactosidase and alkaline phosphatase is presented in Fig. 3. It shows that while the β-galactosidase responds slower than the alkaline phosphatase reporter, it is much more sensitive.

6 In Situ and Online Measurements

An important aspect in biosensing is the ability to monitor in situ and preferably online. For monitoring environmental pollution, having biosensors that can be used in the field dispenses with the need to bring samples to the laboratory, and enables real time assessment of the level of pollution. The electrochemical measurements are highly sensitive, reproducible, and employ a compact analyzer and disposable electrodes. These systems enable simultaneous measurements of several samples. In addition, since the measurement is not optical, it is also possible to perform measurements in crude or turbid solutions – important for determining pollutants in water systems or even in soil.

7 Monitoring Environmental Pollution

Electrochemical biosensors are especially useful for monitoring pollutants such as heavy metals, genotoxic agents, and hydrocarbons [5–7, 30, 36]. It is here that the ability to perform measurements in situ and online is most advantageous. As an example, we constructed a biosensor for cadmium pollution, which consists of the *lacZ* gene expressed under the control of the cadmium-responsive promoter *zntA* [37]. This whole cell biosensor could detect, within minutes, nanomolar concentrations of cadmium in water, seawater, and soil samples, and it was used for continuous online and in situ monitoring [5]. In addition to cadmium, this biosensor can detect the presence of a variety of heavy metals – mercury, zinc, and copper. Moreover, it was possible to monitor the concentration of cadmium for many hours online [5].

8 Conclusions

Electrochemical monitoring of whole cell sensors offers the tools for in situ and online monitoring with simple equipment that can easily be manipulated and miniaturized. Therefore we believe that with the advance of portable computers and electronic devices, electrochemical monitoring will become a method of choice, especially for monitoring environmental pollution.

References

1. Biran I, Klimentiy L, Hengge-Aronis R, Ron EZ, Rishpon J (1999) On-line monitoring of gene expression. Microbiology 145:2129–2133
2. Popovtzer R, Neufeld T, Biran D, Ron EZ, Rishpon J, Shacham-Diamand Y (2005) Novel integrated electrochemical nano-biochip for toxicity detection in water. Nano Lett 5:1023–1027
3. Mittelmann AS, Ron EZ, Rishpon J (2002) Amperometric quantification of total coliforms and specific detection of *Echerichia coli*. Anal Chem 74:903–907
4. Neufeld T, Schwartz-Mittelmann A, Biran D, Ron EZ, Rishpon J (2003) Combined phage typing and amperometric detection of released enzymatic activity for the specific identification and quantification of bacteria. Anal Chem 75:580–585
5. Biran I, Babai R, Levcov K, Rishpon J, Ron EZ (2000) Online and in situ monitoring of environmental pollutants: electrochemical biosensing of cadmium. Environ Microbiol 2:285–290
6. Paitan Y, Biran D, Biran I, Shechter N, Babai R, Rishpon J, Ron EZ (2003) On-line and in situ biosensors for monitoring environmental pollution. Biotechnol Adv 22:27–33
7. Paitan Y, Biran I, Shechter N, Biran D, Rishpon J, Ron EZ (2004) Monitoring aromatic hydrocarbons by whole cell electrochemical biosensors. Anal Biochem 335:175–183
8. Abd-El-Haleem D, Zaki S, Abulhamd A, Elbery H, Abu-Elreesh G (2006) Acinetobacter bioreporter assessing heavy metals toxicity. J Basic Microbiol 46:339–347
9. Bontidean I, Mortari A, Leth S, Brown NL, Karlson U, Larsen MM, Vangronsveld J, Corbisier P, Csoregi E (2004) Biosensors for detection of mercury in contaminated soils. Environ Pollut 131:255–262
10. Brandt KK, Holm PE, Nybroe O (2006) Bioavailability and toxicity of soil particle-associated copper as determined by two bioluminescent Pseudomonas fluorescens biosensor strains. Environ Toxicol Chem 25:1738–1741

11. Chouteau C, Dzyadevych S, Durrieu C, Chovelon JM (2005) A bi-enzymatic whole cell conductometric biosensor for heavy metal ions and pesticides detection in water samples. Biosens Bioelectron 21:273–281
12. Fujimoto H, Wakabayashi M, Yamashiro H, Maeda I, Isoda K, Kondoh M, Kawase M, Miyasaka H, Yagi K (2006) Whole-cell arsenite biosensor using photosynthetic bacterium *Rhodovumul sulfidophilum: Rhodovumul sulfidophilum* as an arsenite biosensor. Appl Microbiol Biotechnol 73:332–338
13. Hakkila K, Green T, Leskinen P, Ivask A, Marks R, Virta M (2004) Detection of bioavailable heavy metals in EILATox-Oregon samples using whole-cell luminescent bacterial sensors in suspension or immobilized onto fibre-optic tips. J Appl Toxicol 24:333–342
14. Shetty RS, Deo SK, Shah P, Sun Y, Rosen BP, Daunert S (2003) Luminescence-based whole-cell-sensing systems for cadmium and lead using genetically engineered bacteria. Anal Bioanal Chem 376:11–17
15. Shetty RS, Deo SK, Liu Y, Daunert S (2004) Fluorescence-based sensing system for copper using genetically engineered living yeast cells. Biotechnol Bioeng 88:664–670
16. Verma N, Singh M (2005) Biosensors for heavy metals. Biometals 18:121–129
17. Campbell DW, Muller C, Reardon KF (2006) Development of a fiber optic enzymatic biosensor for 1, 2-dichloroethane. Biotechnol Lett 28:883–887
18. Leedjarv A, Ivask A, Virta M, Kahru A (2006) Analysis of bioavailable phenols from natural samples by recombinant luminescent bacterial sensors. Chemosphere 64:1910–1919
19. Phoenix P, Keane A, Patel A, Bergeron H, Ghoshal S, Lau PC (2003) Characterization of a new solvent-responsive gene locus in *Pseudomonas putida* F1 and its functionalization as a versatile biosensor. Environ Microbiol 5:1309–1327
20. Tizzard AC, Bergsma JH, Lloyd-Jones G (2006) A resazurin-based biosensor for organic pollutants. Biosens Bioelectron 22:759–763
21. Xu Z, Mulchandani A, Chen W (2003) Detection of benzene, toluene, ethyl benzene, and xylenes (BTEX) using toluene dioxygenase-peroxidase coupling reactions. Biotechnol Prog 19:1812–1815
22. Farre M, Goncalves C, Lacorte S, Barcelo D, Alpendurada MF (2002) Pesticide toxicity assessment using an electrochemical biosensor with *Pseudomonas putida* and a bioluminescence inhibition assay with Vibrio fischeri. Anal Bioanal Chem 373:696–703
23. Huang WE, Wang H, Zheng H, Huang L, Singer AC, Thompson I, Whiteley AS (2005) Chromosomally located gene fusions constructed in Acinetobacter sp. ADP1 for the detection of salicylate. Environ Microbiol 7:1339–1348
24. Lei Y, Mulchandani P, Chen W, Wang J, Mulchandani A (2004) Whole cell-enzyme hybrid amperometric biosensor for direct determination of organophosphorus nerve agents with *p*-nitrophenyl substituent. Biotechnol Bioeng 85:706–713
25. Lei Y, Mulchandani P, Wang J, Chen W, Mulchandani A (2005) Highly sensitive and selective amperometric microbial biosensor for direct determination of *p*-nitrophenyl-substituted organophosphate nerve agents. Environ Sci Technol 39:8853–8857
26. Matsui N, Kaya T, Nagamine K, Yasukawa T, Shiku H, Matsue T (2006) Electrochemical mutagen screening using microbial chip. Biosens Bioelectron 21:1202–1209
27. Norman A, Hansen LH, Sorensen SJ (2006) A flow cytometry-optimized assay using an SOS-green fluorescent protein (SOS-GFP) whole-cell biosensor for the detection of genotoxins in complex environments. Mutat Res 603:164–172
28. Belkin S, Smulski DR, Vollmer AC, Van Dyk TK, LaRossa RA (1996) Oxidative stress detection with *Escherichia coli* harboring a katG'::lux fusion. Appl Environ Microbiol 62:2252–2256
29. Choi JW, Park KW, Lee DB, Lee W, Lee WH (2005) Cell immobilization using self-assembled synthetic oligopeptide and its application to biological toxicity detection using surface plasmon resonance. Biosens Bioelectron 20:2300–2305
30. Neufeld T, Biran D, Popovtzer R, Erez T, Ron EZ, Rishpon J (2006) Genetically engineered *pfabA pfabR* bacteria: an electrochemical whole cell biosensor for detection of water toxicity. Anal Chem 78:4952–4956

31. Sorensen SJ, Burmolle M, Hansen LH (2006) Making bio-sense of toxicity: new developments in whole-cell biosensors. Curr Opin Biotechnol 17:11–16
32. Sørensen J, Nicolaisen MH, Ron E, Simonet P (2009) Molecular tools in rhizosphere microbiology – from single-cell to whole-community analysis. Plant and Soil 321:483–513
33. Belkin S (2003) Microbial whole-cell sensing systems of environmental pollutants. Curr Opin Microbiol 6:206–212
34. Daunert S, Barrett G, Feliciano JS, Shetty RS, Shrestha S, Smith-Spencer W (2000) Genetically engineered whole-cell sensing systems: coupling biological recognition with reporter genes. Chem Rev 100:2705–2738
35. Scott DL, Ramanathan S, Shi W, Rosen BP, Daunert S (1997) Genetically engineered bacteria: electrochemical sensing systems for antimonite and arsenite. Anal Chem 69:16–20
36. Ron EZ (2007) Biosensing environmental pollution. Curr Opin Biotechnol 18:252–256
37. Babai R, Ron EZ (1998) An *Escherichia coli* gene responsive to heavy metals. FEMS Microbiol Lett 167:107–111

Adv Biochem Engin/Biotechnol (2010) 117: 85–108
DOI: 10.1007/10_2009_16

Published online: 13 January 2010

Microbial Cell Arrays

Tal Elad, Jin Hyung Lee, Man Bock Gu, and Shimshon Belkin

Abstract The coming of age of whole-cell biosensors, combined with the continuing advances in array technologies, has prepared the ground for the next step in the evolution of both disciplines – the whole cell array. In the present chapter, we highlight the state-of-the-art in the different disciplines essential for a functional bacterial array. These include the genetic engineering of the biological components, their immobilization in different polymers, technologies for live cell deposition and patterning on different types of solid surfaces, and cellular viability maintenance. Also reviewed are the types of signals emitted by the reporter cell arrays, some of the transduction methodologies for reading these signals, and the mathematical approaches proposed for their analysis. Finally, we review some of the potential applications for bacterial cell arrays, and list the future needs for their maturation: a richer arsenal of high-performance reporter strains, better methodologies for their incorporation into hardware platforms, design of appropriate detection circuits, the continuing development of dedicated algorithms for multiplex signal analysis, and – most importantly – enhanced long term maintenance of viability and activity on the fabricated biochips.

Keywords Whole-cell arrays • Biosensors • Cell immobilization • Cell deposition • Toxicity testing • Gene expression • Bioreporters

---

T. Elad and S. Belkin (✉)
Institute of Life Sciences, The Hebrew University of Jerusalem, Jerusalem 91904, Israel
e-mail: shimshon@vms.huji.ac.il

J.H. Lee
Department of Environmental Science and Engineering, Gwangju Institute of Science and Technology, Gwangju, Republic of Korea

M.B Gu (✉)
College of Life Sciences and Biotechnology, Korea University, Seoul, Republic of Korea
e-mail: mbgu@korea.ac.kr

Contents

1 Introduction

1.1 The Array Concept: Nucleotides, Proteins, and Cells

Within a relatively very short period, biological microarrays have revolutionized our ability to identify, characterize, and quantify biologically relevant molecules. The principle in all arrays is almost identical: a large family of well-defined reactive molecules is fixed onto a mapped solid surface grid, and exposed to a multicomponent analyte mixture. Sites at which a recognition event has occurred (such as by a complementary nucleic acid sequence) are identified by one of several possible detection techniques (e.g., fluorescence). The characteristics of the sample – and hence the constituents and/or the response of the studied system – can then be discerned from the identity and nature of the bioreceptor molecules occupying these sites. Using this principle, an increasingly large number of applications are being developed in medicine, biology, toxicology, drug screening, and more.

Most arrays in use today are based on nucleic acid oligonucleotides [1–3]; other arrays contain antibodies [4], enzymes [5], or other proteins [6]. The advantages in all of these configurations mostly stem from the extremely high specificity inherent in each individual recognition event occurring on the array surface. A different analytical approach may be generated if the array components are not molecules but rather live cells. Although much of the specificity characterizing molecular recognition is lost, this is more than compensated for by the ability to assay directly biological effects on live systems. This approach, initially aimed towards gene expression studies [7], can be successfully utilized for numerous additional applications, efficiently combining effect-testing with analyte identification. Consequently, a variety of future applications are envisaged for such cell-based systems in the medical, industrial, pharmaceutical, and environmental realms, including drug and chemicals screening or environmental monitoring [8–12]. In this chapter we review the state-of-the-art in this rapidly developing field, focusing on microbial whole-cell arrays.

1.2 Advantages of Cell Array Technology

In contrast to biosensing technologies based on molecular recognition, the use of whole cells as sensing entities allows the investigation of *the activity* of the tested sample rather than *the identity* of its components. Functional cellular responses that can be analyzed in this manner include gene expression, metabolic activity, viability, bioavailability, toxicity, and genotoxicity, measuring either specific or global biological effects of the target analyte(s). Most of these vital responses can only be assayed by the use of live systems; while chemical methods may often yield lower detection thresholds and provide a more sensitive analytical performance, no chemical assay can provide information on the effects of the tested compound as sensed by a live cell. Furthermore, the use of live cells allows for reagent-less, nondestructive real-time monitoring of the biological effects as they develop, with no need for preparatory and analytical steps such as staining or hybridization. In an array format, these advantages are supplemented by the combinatorial effects of multiplexed sensors, comparable to the principles inherent in electronic nose [13] or tongue [14] devices. Once the technical hurdles outlined in the following sections are surmounted, a miniaturized array format will permit a high-throughput sample analysis, superior to current microtiter plate-based screening technologies. The envisaged miniaturization can eventually lead to the construction of portable instrumentation for both laboratory and field use, for applications such as toxicity assessment or mutagenicity testing. The rapidly developing field of molecular systems biology could doubtlessly also make use of such tools for the unraveling of complex biological processes.

1.3 Eukaryotic Vs Prokaryotic Cell Arrays

While the current chapter focuses on prokaryotic cell arrays, eukaryotic systems serve as good examples for the growing interest in the field of whole-cell arrays for high-throughput screening. It has been suggested that arrays of eukaryotic cells can be used in a variety of applications, including biosensing, single-cell analysis, therapeutic agent identification, and quantitative cell biology [15–18]. The techniques used for the generation of such arrays and for the positioning of cells vary, ranging from photolithography to inkjet technology [19–21]. Specific examples include patterning of cell-adhesive self-assembled monolayers using microcontact printing [22], photo- and electropatterning of hydrogel-encapsulated living fibroblasts arrays [23], and inkjet printing of viable mammalian cells [24]. Of special weight is the development of a microarray-driven technique for the analysis of gene function in mammalian cells [25]. Upon the addition of adherent cells and a lipid transfection reagent, microscope slides printed with cDNAs become living microarrays, with clusters of cells overexpressing defined gene products. Since its introduction, this "transfected cell microarray" technique was used for the identification of apoptotic genes [26, 27], modified for large-scale RNAi studies [28–30] and offered as a tool for high-throughput drug discovery [31].

Whereas eukaryotic cell arrays, particularly those based on mammalian systems, possess the unique advantage of more closely simulating human cellular responses, prokaryotic cell systems have numerous compensatory benefits. The cells are easy to grow and maintain, large and homogenous populations are readily obtainable, and suitable cell immobilization and preservation methodologies are available. Prokaryotic cells are also more robust and less sensitive to their physical and chemical environment, and less susceptible to biological contamination. Furthermore, microbial cells are much more amenable to the physical and/or chemical manipulations required to pattern them in the array format. Possibly the most important bacterial characteristic in this context is the facility by which they can be genetically tailored to emit the desired signal in the presence of target compound(s) or specific environmental conditions. In most cases, this is achieved by the fusion of a sensing element – a selected promoter – to a suitable molecular reporter system [32–36]. The expression of two independent reporter systems in a single organism has also been reported [37–39]. Furthermore, such sensor cells can detect analytes in different media, such as water, gas, and soil [40–42].

To progress from a panel of genetically engineered sensor cells to an actual array on a solid platform, several biological and physical issues need to be addressed. These are covered in the following sections.

2 Microbial Cell Array Building Blocks

2.1 Genetic Engineering of Microbial Reporter Cells and the Panel Concept

One of the main advantages inherent in the use of microorganisms as building blocks for whole-cell arrays is the facility by which they can be genetically engineered to respond by a dose-dependent signal to environmental stimuli. Two parallel research approaches have been employed for this purpose, focusing on either constitutive or inducible reporter gene expression, often referred to as "lights off" and "lights on" assays [8].

The "lights off" concept is based on measurement of the decrease in signal intensity, such as that produced by a naturally luminescent bacterium (*Vibrio fischeri*) [43] or a genetically modified one [8]. More relevant to the present chapter is the "lights on" approach, based on the molecular fusion of a reporting gene system to gene promoters from selected stress response regulons. It has been demonstrated that with the use of the appropriate stress-responsive promoters it is possible to construct bacterial reporter strains which generate a dose-dependant signal in response, for example, to the presence of heat shock-inducing agents [44], oxidants [32, 34], or membrane-damaging substances [45]. As no single reporter strain is expected to cover all potential cellular stress factors, it has been proposed that a panel of such stress-specific strains be used [33]. Similar panels have been shown to respond sensitively to environmental pollutants such as dioxins [46] and endocrine disruptors [47]. For the development of genotoxicity assays, the promoters serving as the sensing elements were selected from among DNA-repair operons such as the SOS system; the reporters used were either bacterial *lux* or β-galactosidase [36, 48]. Other reports proposed the use of a green fluorescent protein (GFP) gene from the jellyfish *Aquorea victoria*, or its variants, as an alternative reporter system for the same purpose [49, 50]. Using GFP as a reporter, Norman et al. [51] have demonstrated that the ColD plasmid-borne *cda* gene promoter was preferable to other SOS gene promoters *recA*, *sulA*, and *umuDC*. A yeast-based (*Saccharomyces cerevisiae*) GFP system for genotoxicity assessment is being continuously improved upon by Walmsley and coworkers [52–54].

Another class of inducible systems includes those that can sensitively detect a specific chemical or a group of chemicals. They are usually based on promoters of genes involved in the metabolism pathway of, or the resistance mechanism to, the compound to be detected. Since the pioneering work of Sayler and coworkers in the construction of a *lux* fusion for the specific detection of naphthalene and salicylate [55, 56], there has been a steady stream of similar constructs responsive to different organic or inorganic pollutants and classes of pollutants. Bioluminescence has served as the reporter in many of these cases, with a few examples of β-galactosidase activity and GFP accumulation [8, 10, 57–59]. In view of their global significance as environmental pollutants, heavy metals were the targets for a large number of these sensor systems [60].

2.2 Reporters and Signals

As indicated in the preceding section, the signals emitted by microbial reporters can be generated either by the presence of a protein (e.g., GFP or other fluorescent proteins), a carotenoid (*crtA*; [61, 62]), or by the activity of an enzyme. The latter category includes bacterial luciferase (*lux*) and β-galactosidase (*lacZ*), as well as others: insect luciferase (*luc*), alkaline phosphatase (*phoA*), β-glucuronidase (*uidA*), and β-lactamase (*bla*) [63–67]. Depending on the reporter gene used and on the substrate provided to its product, the emitted signals can be detected optically (by bioluminescence, fluorescence, or colorimetry), or electrochemically [57]. Additional assays, also applicable to array formats, may be based on cell viability, for example, by the Live/Dead system [68] or by surface plasmon resonance analysis [69], cell length [70], or cellular well-being and growth. The latter has been demonstrated microscopically, using single YFP-tagged cells in narrow microfluidic channels [71].

2.3 Data Analysis: Interpreting the Array Response Pattern

Data emanating from the individual responses of an array of specific stress-responsive bacteria should not only allow the detection of a wide range of toxic chemicals but also indicate the type of biological activity involved. With a sufficient number of array members, each chemical or group of chemicals will be characterized by its own specific signature, which can then be used both to identify the chemicals in the sample and to indicate their biological activity.

Indeed, Lee et al. [72] demonstrated the different response patterns of a cell array chip constructed from 20 luminescent reporter strains following a separate exposure to three chemicals of different biological activities. The same group also reported the fabrication of an oxidative stress-specific bacterial cell array chip [73]. Nine chemicals were selected to test the capabilities of this array when analyzing different oxidative effects. The results clearly show that each of the toxicants tested generated a distinct response pattern. Moreover, it was shown that each strain was responsive to one or more of the compounds tested, but that the responses were dependent upon the production of superoxide radicals, i.e., the strains were unresponsive to compounds that were similar in structure but lacked the ability to generate such radicals. These findings demonstrate that such an array can be used to elucidate the nature of the adverse effect of toxic chemicals, and suggest that their response pattern-based classification should be made according to the chemicals' mode of action rather than their structure.

An attempt to identify toxicants on the basis of their biologic fingerprint by a pattern classification algorithm was made by Ben-Israel et al. [74], using *Escherichia coli* strains carrying *lux* genes fused to several stress-responsive gene promoters (*micF*, *lon*, *fabA*, *katG*, and *uspA*). The reporter strains were exposed to various toxic chemicals, and the recorded luminescence was used for the characterization of the signature of each toxicant. A compound's signature was determined on the basis of the dependence between the induced luminescence and the compound's concentration. The data were analyzed with SAS software (SAS Institute Inc., Cary, N.C.) by use of a discriminant analysis (DA) method, allowing quantitative and qualitative assessment of toxic chemicals tested. Of the 25 compounds tested, 23 were identified by this strategy in a 3-h procedure. The signature of a binary mixture was predicted by use of the learning data characterizing each toxicant separately and a good correlation ($R^2 \geq 0.85$) was found between the observed and the predicted response patterns. The authors also showed that closely related chemicals, in particular those with a similar mode of action, tended to cluster in the same subgroup.

The results of Lee et al. [73] and Ben-Israel et al. [74] imply that the biological recognition strategy can provide a means for assaying a compound's mode of action. Maybe the strongest evidence for the validity of this hypothesis was provided when hierarchical clustering analysis was applied to a compendium of drug-hypersensitivity profiles generated by screening dozens of different compounds against a comprehensive collection of yeast deletion mutants [75, 76]. Hierarchical clustering is widely used for DNA microarray expression data interpretation [1, 77], and was demonstrated to be a handy tool for the identification of the cellular functions of uncharacterized open reading frames (ORFs) via a reference database of distinct expression profiles [78]. When hierarchical clustering was applied to the compendium of sensitivity profiles mentioned above, compounds with similar cellular effects clustered together, revealing anticipated and novel insights into their mode of action.

To evaluate further options for array data analysis we have generated a very large data set using five bioluminescent reporter strains exposed to five model toxicants and to a buffer control. Forty randomly arrayed repeats were carried out in 384-well microtiter plates, and the emitted light was quantified every 5 min for 2 h [79]. The data were then analyzed using different mathematical and statistical approaches. Five of the six treatments were identified by an artificial neuron network 30 and 60 min after exposure, while all six were identified by the same method after 120 min. Bayesian decision theory and the nonparametric nearest-neighbor technique [80] were also applied to the collected data [81]. Similarly, classifiers were designed based on the data collected 30, 60, and 120 min after exposure. The Bayesian classifiers performed better and were able to identify the sample's contents within 30 min with an error rate estimate that did not exceed 3% at a 95% confidence level and with zero false negatives. To test the validity of the Bayesian pattern classification algorithm in real-world environments, tap water and wastewater samples were spiked with two of the model chemicals (potassium cyanide and paraquat). Subsequently, the spiked samples were introduced to the reporter strain panel, the response patterns of which were correctly recognized by the Bayesian classifier [81].

2.4 Cell Array Substrates

Similarly to oligonucleotide or protein arrays, cell arrays need to be spotted on a compatible substrate. Numerous materials are available for this purpose, including silicon, glass, and various polymers. Silicon, extensively used within the semiconductor field, is an attractive option, as integrated circuit technologies can easily be employed for cell array fabrication [82, 83]. Micromachining technologies make it possible to tailor the silicon chip to the required topographical specifications, by patterning microfluidic channels, microchambers, valves, and additional structures

[84, 85]. Cell arrays might also be generated directly on electrical components, such as photo-diodes, light emitting diodes, or field effect transistors.

Although well-established, the machining of silicon platforms is relatively complex, time-consuming, and expensive. Glass, being highly biocompatible and transparent, is an attractive and cost-effective alternative. Similarly to silicon, it is amenable to the etching of microfluidic channels and chambers [86], as well as to various chemical or physical treatments that modify surface characteristics and cell attachment [87–89].

Various polymers are widely used to assemble microfluidic channels and other surface structures. Polydimethylsiloxane (PDMS) is often preferred as an array material because of its ease in handling, optical transparency, and biocompatability. Tani et al. [90] reported a three-dimensional microfluidic network system for constructing on-chip bacterial cell array bioassays; microchannels fabricated on two separate PDMS layers were connected via perforated microwells on the silicon chip to form a three-dimensional microfluidic network. A PDMS maze was used to observe bacterial motion under nutrient depletion to study quorum formation [91]. PDMS channels have been similarly used to study the persistence of bacteria under stress [71], where the proliferation of single cells could be monitored.

Other substrates reported for the support of cell arrays included optical fiber bundles [92, 93] and gold surfaces [69]. In the former case, the responses from individual cells could be measured independently. In the latter, cells were immobilized on gold plates using self-assembled cysteine-terminated synthetic oligopeptides. An interesting tack towards a broad use of cell array biochips was pursued by Ingham et al. [94]. The researchers have fabricated a miniaturized, disposable microbial culture chip, a "micropetri dish", by microengineering growth compartments on top of porous aluminum oxide (PAO). The chip, placed on nutrient agar, acts as the surface on which an exceptional number of microbial samples (up to one million wells per 8 $\times$ 36 mm chip) can be grown, assayed, and recovered.

2.5 Cell Array Deposition Techniques

Several approaches have been proposed and demonstrated for arraying the cells in the required pattern. Useful tools for this purpose are array robots of various kinds. For example, using an Affymetrix 417 arrayer robot (Affymetrix, Santa Clara, CA), Fesenko et al. [95] have fabricated an acrylamide-based hydrogel bacterial microchip (HBMChip). The microchip consisted of an array of hemispherical gel elements, 0.3–60 nL in volume, attached to a hydrophobic glass surface and containing microbial cells. Whilst array robots may be the "natural" choice for cell deposition, others have suggested innovative printing techniques based on ink-jet, laser, or microcontact technologies.

Flickinger et al. [96] formulated reactive microbial inks; piezo tips were used to generate ink-jet deposited *merR::lux*-harboring *E. coli* dot arrays using a latex ink formulation. Luminescence was induced with 20 ng mL^{-1} Hg^{2+}, and the array responded within 1 h. Mercury was also detected with similar immobilized or nonimmobilized constructs at lower concentrations, down to the parts per trillion scale, albeit not in an array format [97–99].

Boland and coworkers developed a method for fabricating bacterial colony arrays on soy agar using commercially available ink-jet printers [100]. Bacterial colony arrays with a density of 100 colonies per cm^2 were obtained by directly ejecting *E. coli* onto soy agar-coated substrates at a high arraying speed. Adjusting the concentration of bacterial suspensions allowed single colonies to be obtained. Barron et al. [101] used a laser-based printing method to transfer genetically modified bioreporters bacteria onto agar-coated slides. Both authors imply that the developed technique enables an array consisting of different bacterial strains to be fabricated. The former suggests loading different bacteria strains into different cartridges for creating mixed colony arrays from different bacteria types. The latter indicates that, as an orifice-free technique, it allows the patterning of different cell types adjacently by switching the laser's target disks without concerns relating to cleaning cycles or potential contamination.

Another recently proposed methodology for the construction of bacterial arrays is microcontact printing. The utilization of PDMS stamps for printing an *E. coli* array on agarose was demonstrated by Xu et al. [102]. Taking a somewhat different approach, Weibel et al. [103] described the use of microcontact agarose stamps prepared by molding against PDMS masters to print bacterial colony arrays on agar plates. By this technique, the group has generated a colony array out of several different strains of bacteria in a scale of 1–2 mm per element. Alternatively, microcontact printing of an adhesive organic monolayer was used in a four-step soft lithography process to fabricate 12 μm square bacteria "corrals" on a silicon wafer substrate [104].

2.6 Cell Immobilization, Maintenance, and Long-Term Storage

The live cell "spots" on the array surface need to be deposited in such a manner that will not only place the cells in the appropriate pattern in relation to each other and to the sensing device's signal transducer, but will also allow long-term cell preservation. Various approaches for viability and activity maintenance of live reporter cells over prolonged periods of time for environmental monitoring and toxicity assessment have been reviewed by Bjerketorp et al. [105]. Reported solutions include freeze/vacuum drying [106–108] as well as cell encapsulation and entrapment in a large variety of polymers. Methods used for the fabrication of bacterial cell arrays (mostly of genetically engineered bioluminescent reporters) included agar [72, 73], agarose [90], alginate [109, 110], carrageenan [111], collagen [112], latex [96], and polyacrylamide [95]. The latter has been reported to cause loss of viability [109, 113]. Noteworthy is the work of Akselrod et al. [114] who, following the pioneering work of Ashkin and coworkers in optical force-based particle manipulation [115,

116], have assembled microarrays of living bacterial cells in a polyethylene glycol diacrylate (PEGDA) hydrogel with optical traps. Using PEGDA as a scaffold to support the optically organized arrays and fix the position of the cells, a 5 $\times$ 5 two-dimensional array of *E. coli* was formed in the hydrogel and cell viability after 43 h was confirmed by *gfp* induction. An entirely different approach to on-chip long-term cell viability maintenance can be inferred from Balagaddé et al. [117]. A chip-based bioreactor that uses microfluidic plumbing networks to prevent actively biofilm formation was created. The device allowed steady-state growth in six independent 16-nL reactors which served as "microchemostats" and enabled long-term culture maintenance.

Only a few of the reports referred to above have investigated long-term cell viability maintenance or the ability to store the fabricated array for long periods. Recombinant bioluminescent *E. coli* responsive to nalidixic acid and laser-printed on agar-coated slides maintained their activity after shipment at ambient temperature followed by storage for up to 2 weeks at 4°C [101]. Furthermore, it was indicated that the orifice-free aspect of this laser procedure may be useful in the design of an off-the-shelf bacterial biosensor that can be stored without loss of activity, since it allows for transfer of lyophilized bacteria. The active sensor lifetime and the shelf lifetime of an optical imaging fiber-based live bacterial cell array biosensor were investigated as well [93]. The sensors retained their sensing ability for at least 6 h when stored in an ambient environment and demonstrated a shelf life of 2 weeks at 4°C.

Regardless of the inherent analytical qualities of any live cell array and of its performance when freshly deposited, future implementations of such arrays will depend on successful long-term storage and viability preservation. With this objective in mind, the results of the studies summarized in this section, while providing an improvement over earlier reports, are nevertheless still unsatisfactory. Future studies will need to address this essential aspect of any future whole-cell sensor array; possibly one of the more promising avenues of research will focus on the synthesis and formulations of new immobilization matrices [105]. Another potentially useful approach may be in the testing of new osmo- and cryo-protectants, including novel compounds isolated from highly desiccation- or freeze-resistant organisms.

2.7 Enhancement of Array Sensitivity

One of the drawbacks inherent in the array format is that the signal emanating from the small amount of the cells that can be concentrated in microliter- or even nanoliter-size spots may be very low. An increase in signal intensity would allow more sensitive analyte detection, as well as the use of simpler (and thus cheaper) detectors.

Enhancement of signal intensity, or modifications in response sensitivity and the timing of its onset, can be achieved by genetic manipulations of the promoter region, the reporter gene(s), or the host cell. Insights as to possible avenues for modifying reporter specificities were provided by Galvao and de Lorenzo [118]; considerations for performance optimization were elegantly outlined by van der Meer et al. [119] as well as by Marqués et al. [120]. An enhancement of biosensor capability by modifying the origin of reporter genes was also demonstrated [121]. For example, *E. coli* strains carrying fusions of selected oxidative stress-responsive promoters to *Photorhabdus luminescens lux* showed higher bioluminescent levels than strains carrying the same promoters fused to the *luxCDABE* genes from *V. fischeri*, while the detection thresholds of the strains were similar, regardless of the luciferase used. A substrate-specific approach to electrochemical signal amplification was reported by Neufeld et al. [122]; *p*-aminophenol (PAP), the end-product of the activity of the β-galactosidase reporter enzyme, acted as an activator of the sensor element used in that construct, *fabA*.

A complementary tactic for sensitivity enhancement may lay in improving optical signal acquisition rather than reporter performance. Reporter response monitoring, down to single-molecule detection, was achieved using laser-induced fluorescence coupled with confocal spectroscopy [123]. Also relevant in this field are efforts to develop complementary metal oxide semiconductor (CMOS) detector modules for bioluminescence and fluorescence measurements, as well as the use of fluorescence flow cytometry (FCM) to measure fluorescent reporter response [124].

3 Microbial Cell Array Applications

3.1 Toxicity Assessment and Monitoring

While the envisaged applications of live whole-cell arrays are numerous, many of the current reports either address environmental applications or test the effect of environmentally relevant chemicals. Most prominent among these are the attempts to use the arrays as tools in toxicity testing. Standard approaches to toxicity bioassays center around the quantification of the negative effects of the tested sample on a test organism population. Originally based on the use of live organisms, recent years have seen a shift to cellular and subcellular alternatives; particularly attractive in this respect is the possibility of genetically tailoring microorganisms to respond to specific sets of toxic chemicals [8]. Microbial cell arrays are a logical step forward: a panel of genetically engineered microorganisms, each modified to respond to a different class of chemicals, and together covering a broad range of potential toxic effects. In each of the panel members, different gene cascades are elicited in response to different stress factors and different biochemical responses are expressed [125–128]. The use of such a strain panel for multiplexed toxicity analysis has been proposed by several authors [33, 58, 129, 130]. Published reports on multigenotype cell arrays designated for toxicity assessment purposes are reviewed hereafter. These, as well as multigenotype cell arrays designated for other purposes, are summarized in Table 1.

Table 1 Multigenotype microbial cell array implementations

Purpose	Reporter	Signal	Substrate	Immobilization matrix/method	Reference
Toxicity assessment and monitoring	*E. coli*	Fluorescence	Fiber optic	–	[93]
	E. coli	Bioluminescence	Silicon	Agarose	[90]
	E. coli	Bioluminescence	Plastic	Agar	[72, 73]
	E. coli	Electrical current	Silicon	–	[132]
Gene expression and function analysis	*E. coli*	Bioluminescence	Nylon membrane	–	[7]
	S. cerevisiae	Growth	Agar plates	–	[135]
	E. coli	Growth	Agar plates	–	[145]
Antibody detection	GP/GN* pathogens	Fluorescence	Nitrocellulose-coated glass	Adsorption	[89]
Protein detection	*E. coli*	Fluorescence	Glass	DEP	[148]
Sugar determination	*E. coli*	Electrical current	Silicon	κ-Carrageenan	[111]

*Gram positive/Gram negative

Biran and Walt [92] introduced a high-density ordered array of individually addressed single cells, occupying thousands of microwells etched on the distal end of an optical imaging fiber. While no surface modification was needed in order to place and maintain yeast cells in the microwells, for fabricating *E. coli* arrays the microwell surface was modified with polyethylenimine (PEI) to allow better adhesion of the cells to the microwells through electrostatic interactions. In two subsequent studies, Walt and coworkers have detected mercury [131] and genotoxins [93] using such cell arrays. In the first study, a genetically modified *E. coli* strain, containing the *lacZ* reporter gene fused to the heavy metal-responsive gene promoter *zntA*, was used to fabricate a mercury biosensor. Single-cell *lacZ* expression was measured when the array was exposed to mercury and a response to 20 ng mL^{-1} Hg^{2+} could be detected after 1 h. In the second study, mitomycin C at similar concentrations was detected by *E. coli* cells carrying a *recA::gfp* fusion after 90 min. The researchers further demonstrated the simultaneous embedding of different strains into the array. As cells were randomly dispersed into the microwells by centrifugation, each strain was encoded with a unique fluorescent dye or protein. This fluorescent signature, in turn, facilitated the locating of strain members [92].

Tani et al. [90] presented an on-chip format for high-throughput whole-cell bioassays (Fig. 1). Their assembly included a silicon substrate, perforated with wells and placed between two microchanneled PDMS layers. One PDMS layer was used to inject a reporter-agarose mix into the wells. After gelation, the other PDMS layer was placed on the other side of the silicon substrate and was used to expose the immobilized reporter cells to the sample. Using this setup, the interactions between various types of samples and strains could be monitored in one assembly in a combinatorial fashion. The operation of the array was exemplified by the simultaneous exposure of different *E. coli* hosts, all harboring a plasmid-borne *umuD::luc* fusion, to different concentrations of mitomycin C. The toxin was detected at concentrations down to 0.02 mg L^{-1}.

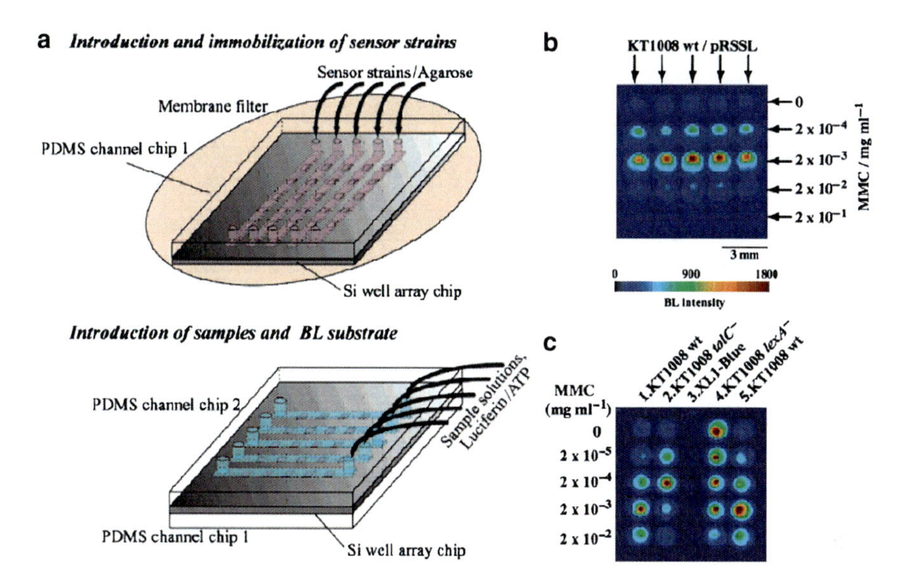

Fig. 1 A whole-cell array combined with a three-dimensional microfluidic network. (**a**) Agarose immobilization of the reporter cells on a PDMS chip and sample introduction. (**b**) Bioluminescence imaging using a bacterial reporter strain harboring a plasmid-borne *umuD::luc* fusion, in response to mitomycin C. (**c**) Bioluminescence imaging using several reporter strains in response to mitomycin C. From Tani et al. [90] with permission

Lee et al. [72] used a standard 384-well plate and a self-constructed 96-well acryl chip as platforms for the development of two biosensor arrays. Twenty recombinant bioluminescent bacteria, harboring different promoters fused to bacterial *lux* genes, were deposited in the wells of either the chip or the 384-well plate after agar immobilization and the responses from the cell arrays were characterized using three chemicals that cause either superoxide damage (paraquat), DNA damage (mitomycin C), or protein/membrane damage (salicylic acid); only 2 h were needed for analysis. On the same acryl chip platform, an oxidative stress-specific cell array was fabricated [73]. The chip consisted of 12 agar-immobilized bioluminescent strains, each responsive to a different type of oxidative stress. Array performance, tested with nine chemicals, displayed the desired selectivity: not only did the array respond to paraquat and four of its radical producing structural analogs, but also exhibited different response patterns to each of the five substances.

The use of an electrode platform for toxicity detection was demonstrated by Popovtzer et al. [132], who constructed an electrochemical biochip for water toxicity detection. The silicon biochip contained an array of 100-nL electrochemical chambers, harboring *E. coli* cells, genetically engineered to express *lacZ* in response to stress promoter (*grpE* and *dnaK*) activation. β-Galactosidase activity was monitored by electrochemical determination of the concentration of PAP, the enzymatic hydrolysis product of *p*-aminophenyl β-D-galactopyranoside (PAPG), and a clear electrical signal was produced following an exposure to ethanol (0.5%) or phenol

(1.6 mg L^{-1}). This approach, while requiring the addition of an external substrate, does not involve lysis of the cells. Like bioluminescence, it is thus suitable for continuous online measurement of enzymatic activity, even in turbid solutions and under anaerobic conditions. The use of an electrochemical signal also appealed to Matsui and colleagues [112], who performed an electrochemical mutagen screening on a microbial chip combined with a scanning electrochemical microscopy (SECM) device. The microbial chip was fabricated by embedding 5 nL of collagen-immobilized *umuC::lacZ Salmonella typhimurium* in microcavities on a glass substrate. Overall, lower limits of detection of 2-aminoflouren, mitomycin C, and 2-aminoanthracene were obtained by the microbial chip as compared to the conventional *umu* test [133], but it should be noted that the definition of the detection limit was different and exposure times were longer.

3.2 Gene Expression and Function Analysis

As indicated in the introductory section, DNA microarrays have been extensively used in recent years for the determination of gene expression and function. Nonetheless, the technology is not free of limitations. Bacterial mRNA isolation might yield undesired artifacts due, for example, to the instability inherent in the molecule [134]. In addition post-transcriptional and post-translational regulations are not detected by transcriptional analysis. Hence, from a function analysis point of view, DNA microarray technology may generate hypotheses rather than establish affiliations between genes to their functions. Live cell arrays offer a solution to these drawbacks, as they dismiss the need for mRNA isolation and keep gene functions in their full cellular context.

For gene expression analysis, Van Dyk et al. [7] – among the first to advance the concept of using live cells as array components – described the LuxArray: a collection of selected 689 nonredundant functional promoter fusions to *P. luminescens luxCDABE* in live *E. coli* strains, representing close to 30% of the predicted transcriptional units in this bacterium. For array manufacturing, duplicate cultures of the *E. coli* reporter strains were grown overnight in a set of 96-well plates. Sterilized 8 × 12 cm porous nylon membranes were placed in contact with solid LB growth media in a culture dish. Printing of spots on top of the membranes at a density of 16 spots per cm^2 was accomplished using a robot equipped with a high-density replication tool. Following cell printing, the arrays were incubated for 6 h at 37°C to allow the cells to grow and to increase the bioluminescent signal. Then the membranes were moved to new plates, containing either LB media or LB media supplemented with nalidixic acid, which were placed at 37°C to continue growth. Images were collected every 2 h from 0 to 8 h after relocation with a cooled charge-coupled device camera. Using this format, new LexA-regulated SOS genes were discovered, as well as new nalidixic acid-upregulated genes that are not a part of the general DNA-damage response.

Other prominent work is that of Tong et al. [135], who introduced synthetic genetic array (SGA) analysis for mapping genetic interaction networks in yeasts. In its simplest form, SGA involves a series of replica-pinning procedures in which mating and meiotic recombination are used to cross a query mutation to a large collection of gene-deletion mutants. The final pinning procedure results in a high density ordered array of double-mutant haploid strains the growth of which is monitored, and nonviable or fitness-reduced double mutants identify functional relationships between genes. When put into practice for the first time [135], 8 query mutations were crossed to a deletion mutant library of all ~5000 nonessential yeast genes [136], and a network of 291 interactions among 204 genes was generated. The authors extrapolated that an order of 300 SGA screens covering selected query genes will provide an effective working genetic scaffold, and in successive work preformed 132 such screens, focused on query genes involved in diverse molecular mechanisms [137]. The large scale analysis yielded a detailed network of connected gene functions and clusters of genetic interaction profiles which, among others, highlighted particular pathways that buffer one another and helped unravel the biological function of uncharacterized genes. Recent applications of SGA analysis provided evidence for interactions between mRNA export factors and sites of mRNA turnover and storage [138], identified gene silencing processes [139], and elucidated regulatory systems of protein function [140]. In addition, SGA analysis permits one to cross query mutations to yeast strain collections other than the collection of viable gene-deletion mutants. Kinase targets and previously uncharacterized essential genes were identified when crossing query mutations to libraries of overexpressed and essential yeast genes, respectively [141, 142].

As illustrated by SGA analysis, the complete set of *S. cerevisiae* deletion mutants [136] has opened the door for high-throughput screens of the yeast genome. Sensitivity tests using ordered arrays of viable gene-deletion mutants at a density of 768 strains per plate have revealed new DNA damage response genes as well as a large group of genes required for multidrug resistance [75, 143]. Moreover, such sensitivity tests were shown to be a powerful tool for inferring the mode-of-action of bioactive compounds [75, 76]. In an attempt to miniaturize further the technology, Xie et al. [144] have developed a strategy for arraying the full collection of ~6,000 yeast deletion strains on a single plate. Their strategy relies on nanoliter cell printing using a microarray robot, followed by cell array imaging using a charge-coupled device system.

The path of exploring genetic interaction networks with large collections of double-mutant strains was recently pursued by bacteriologists as well [145]. The method is termed GIANT-coli and allows mass generation of *E. coli* double mutant strains. It is based on the Hfr conjugation gene transfer system and uses two comprehensive *E. coli* mutant libraries of ~4,000 single-gene deletions, each marked with a different antibiotic-resistance gene. In essence, an Hfr donor strain carrying a marker replacing a selected open reading frame is spread as a lawn on agar plates and mated with robotically arrayed F^- single-gene knockouts carrying a different marker. Finally, cells are pinned onto a plate containing both antibiotics to select for double recombinants, and after growing them for an experimentally determined time, sick and healthy mutants are screened for and scored. When validated with a 12 x 12 genetic interaction matrix – yielding 66 distinct pair-wise double mutant strains – GIANT-coli identified 12 negative or positive synthetic interactions. All identified interactions were either previously described or independently confirmed by established methodologies, affirming the effectiveness of the method.

3.3 Antibody and Protein Detection

Solid phase immunoassays with intact bacterial cells immobilized as antigens on enzyme-linked immunosorbent assay (ELISA) plates have been developed with various bacteria [146]. Utilizing cell surface antigen–antibody interactions, whole-cell assays are particularly useful when the antigens that react with the antibodies against a certain pathogen are unknown. In addition, isolation of pure antigens is often complex and labor intensive [147]. Combining the advantages of a cell-based immunoassay with the benefits of a miniaturized array format, Thirumalapura et al. [89] applied a bacterial array biochip to antibody detection. Gram-negative and Gram-positive strains were printed on nitrocellulose-coated glass substrates with a microarray printer. After cell printing and adsorption, the binding of monoclonal antibodies to surface antigens of the immobilized bacteria was examined with fluorophore-labeled secondary antibodies. Monoclonal antibodies specific for the O-antigen of Gram-negative *Francisella tularensis* or Gram-negative *S. typhimurium* bound to *F. tularensis* or *S. typhimurium*, respectively, but not to other Gram-negative bacteria printed on the same slide. Likewise, Monoclonal antibodies specific for the peptidoglycan of Gram-positive *Staphylococcus aureus* recognized *S. aureus* cells but not Gram-positives *Streptococcus pyogenes* and *Listeria monocytogenes*, while monoclonal antibodies against *L. monocytogenes* specifically bound *L. monocytogenes* cells. Sensitivity was also investigated and was found to be 0.1 $\mu g\ mL^{-1}$. Finally, the authors reported the cell array's successful detection of anti-*F. tularensis* antibodies in canine serum samples declared positive for tularemia.

Cell surface characteristics can also be employed for protein detection. An array biochip was developed in which recombinant bacteria expressing specific surface capture proteins were spatially arrayed in microfluidic channels by dielectrophoresis to detect protein molecules that are difficult to purify and immobilize [148]. Specifically, *E. coli* cells were genetically engineered to express an outer membrane protein containing a streptavidin-binding peptide. Using flow cytometry and fluorescently-labeled streptavidin molecules, the affinity of the capture ligand for the target protein was confirmed. Then "probe cells" and negative control cells, which do not bind to the target, were spatially organized and immobilized by dielectrophoresis on an integrated microfluidic system. The system coupled a glass wafer, on which an electrode array was fabricated, with a PDMS layer, which contained microfluidic channels and was placed on top of the glass substrate. Using two different channels, probe cells were immobilized on one electrode, while negative control cells were immobilized on another. When fluorescently-labeled streptavidin molecules were injected through both channels, a significantly higher fluorescent signal was recorded on the probe cell electrode, demonstrating the protein detection capabilities of the described cell array.

3.4 Other Applications

Cell-based arrays were also proposed for additional uses. Held et al. [111] described a microbial biosensor array electrode platform, integrated in a flow-injection system, for mono- and disaccharide determination. Transport mutants of *E. coli* were immobilized in carrageenan in front of an O_2-sensing gold electrode; selectivity was endowed by the specific sensitivity of each mutant to a different sugar. Sakaguchi et al. [110] constructed a biochemical oxygen demand (BOD) measurement biochip. Nine cavities, 0.7 mm in diameter, were fabricated in a 3×3 array on an acrylic platform. Cells of the marine luminous bacterium *Photobacterium phosphoreum* were immobilized inside the micro-holes using sodium alginate. Samples were introduced into the bacteria-containing micro-holes, in 5-μL aliquots, and the array chip was incubated at room temperature for 20 min. After incubation, the degree of luminescence increase, resulting from cellular assimilation of organic substances, served for BOD quantification. Light intensity was measured using either laboratory equipment or a self-developed on-site system consisting of a digital camera and a laptop computer. Good performance of both systems was reported with standard solutions as well as actual wastewater samples.

4 Concluding Remarks and Future Outlook

Recent advances in array technologies on the one hand, and the coming of age of whole-cell biosensors on the other, have prepared the ground for the next step in the evolution of both disciplines – the whole-cell array. As indicated in the previous section, envisaged applications for such arrays are numerous; nevertheless, it should be clearly stated that to date these applications have yet to be implemented outside the realms of the research laboratory. The present review, which highlights the state-of-the-art in different disciplines essential for a functional microbial array, also serves to bring forward the numerous hurdles which need to be passed before the technology matures. Possibly the most urgent need is to improve dramatically maintenance of cell activity and viability over prolonged periods after array fabrication; this challenge has hardly been addressed to date. Also essential are an improved arsenal of reporter strains, better methodologies for incorporation of such cells into the hardware platforms, development of appropriate detection circuits, and the availability of dedicated algorithms for multiplex signal analysis.

As the paths to all of these objectives are relatively straightforward, it is tempting to envisage how, within a relatively short period, microbial cell arrays may turn into efficient tools for basic microbiological studies as well as for industrial high-throughput chemical/pharmaceutical screening applications, environmental monitoring and food safety.

Acknowledgements M.B. Gu and S. Belkin acknowledge with gratitude the joint research grant in Nano-Science Technology, awarded by the Korean and Israeli Ministries of Science and Technology in the framework of bilateral science and technology cooperation between the two countries. S. Belkin also acknowledges DARPA grant number N00173-01-1-G009, EU 6th framework project Toxichip and a Hebrew University Applied Grant.

References

1. Ehrenreich A (2006) DNA microarray technology for the microbiologist: an overview. Appl Microbiol Biotechnol 73:255–273
2. Lee NH, Saeed AI (2007) Microarrays: an overview. Methods Mol Biol 353:265–300
3. Petersen DW, Kawasaki ES (2007) Manufacturing of microarrays. Adv Exp Med Biol 593:1–11
4. Wingren C, Borrebaeck CAK (2006) Antibody microarrays: current status and key technological advances. OMICS 10:411–427
5. Reymond JL, Babiak P (2007) Screening systems. Adv Biochem Eng Biotechnol 105:31–58
6. Kricka LJ, Master SR, Joos TO, Fortina P (2006) Current perspectives in protein array technology. Ann Clin Biochem 43:457–467
7. Van Dyk TK, DeRose EJ, Gonye GE (2001) LuxArray, a high-density, genomewide transcription analysis of *Escherichia coli* using bioluminescent reporter strains. J Bacteriol 183:5496–5505
8. Belkin S (2003) Microbial whole-cell sensing systems of environmental pollutants. Curr Opin Microbiol 6:206–212
9. Daunert S, Barrett G, Feliciano JS, Shetty RS, Shrestha S, Smith-Spencer W (2000) Genetically engineered whole-cell sensing systems: coupling biological recognition with reporter genes. Chem Rev 100:2705–2738
10. Gu MB, Mitchell R, Kim BC (2004) Whole-cell-based biosensors for environmental biomonitoring and application. Adv Biochem Eng Biotechnol 87:269–305
11. Sørensen SJ, Burmølle M, Hansen LH (2006) Making bio-sense of toxicity: new developments in whole-cell biosensors. Curr Opin Biotechnol 17:11–16
12. Wang P, Xu G, Qin L, Xu Y, Li Y, Li R (2005) Cell-based biosensors and its application in biomedicine. Sens Actuators B Chem 108:576–584
13. Thaler ER, Hanson CW (2005) Medical applications of electronic nose technology. Expert Rev Med Devices 2:559–566
14. Vlasov Y, Legin A, Rudnitskaya A (2002) Electronic tongues and their analytical application. Anal Bioanal Chem 373:136–146
15. Koh WG, Itle LJ, Pishko MV (2003) Molding of hydrogel microstructures to create multiphenotype cell microarrays. Anal Chem 75:5783–5789
16. Lee PJ, Hung PJ, Rao VM, Lee LP (2006) Nanoliter scale microbioreactor array for quantitative cell biology. Biotechnol Bioeng 94:5–14
17. Palková Z, Váchová L, Valer M, Preckel T (2004) Single-cell analysis of yeast, mammalian cells, and fungal spores with a microfluidic pressure-driven chip-based system. Cytom Part A 59:246–253

18. Waterworth A, Hanby A, Speirs V (2005) A novel cell array technique for high-throughput, cell-based analysis. In Vitro Cell Dev Biol Anim 41:185–187
19. Alper J (2004) Biology and the inkjets. Science 305:1895
20. Beske OE, Goldbard S (2002) High-throughput cell analysis using multiplexed array technologies. Drug Discov Today 7:S131–S135
21. Park TH, Shuler ML (2003) Integration of cell culture and microfabrication technology. Biotechnol Prog 19:243–253
22. Mrksich M, Whitesides GM (1995) Patterning self-assembled monolayers using microcontact printing: a new technology for biosensors? Trends Biotechnol 13:228–235
23. Albrecht DR, Tsang VL, Sah RL, Bhatia SN (2005) Photo- and electropatterning of hydrogel-encapsulated living cell arrays. Lab Chip 5:111–118
24. Xu T, Jin J, Cassie G, Hickman JJ, Boland T (2005) Inkjet printing of viable mammalian cells. Biomaterials 26:93–99
25. Ziauddin J, Sabatini DM (2001) Microarrays of cells expressing defined cDNAs. Nature 411:107–110
26. Mannherz O, Mertens D, Hahn M, Lichter P (2006) Functional screening for proapoptotic genes by reverse transfection cell array technology. Genomics 87:665–672
27. Palmer E, Miller A, Freeman T (2006) Identification and characterisation of human apoptosis inducing proteins using cell-based transfection microarrays and expression analysis. BMC Genomics 7:145
28. Kumar R, Conklin DS, Mittal V (2003) High-throughput selection of effective RNAi probes for gene silencing. Genome Res 13:2333–2340
29. Silva JM, Mizuno H, Brady A, Lucito R, Hannon GJ (2004) RNA interference microarrays: high-throughput loss-of-function genetics in mammalian cells. Proc Natl Acad Sci USA 101:6548–6552
30. Wheeler DB, Carpenter AE, Sabatini DM (2005) Cell microarrays and RNA interference chip away at gene function. Nat Genet Suppl 37:S25–S30
31. Bailey SN, Wu RZ, Sabatini DM (2002) Applications of transfected cell microarrays in high-throughput drug discovery. Drug Discov Today 7:S113–S118
32. Belkin S, Smulski DR, Vollmer AC, Van Dyk TK, LaRossa RA (1996) Oxidative stress detection with *Escherichia coli* harboring a *katG'::lux* fusion. Appl Environ Microbiol 62:2252–2256
33. Belkin S, Smulski DR, Dadon S, Vollmer AC, Van Dyk TK, LaRossa RA (1997) A panel of stress-responsive luminous bacteria for the detection of selected classes of toxicants. Water Res 31:3009–3016
34. Lee HJ, Gu MB (2003) Construction of a *sodA::luxCDABE* fusion *Escherichia coli*: comparison with a *katG* fusion strain through their responses to oxidative stresses. Appl Microbiol Biotechnol 60:577–580
35. Min J, Gu MB (2003) Genotoxicity assay using a chromosomally-integrated bacterial *recA promoter::lux* fusion. J Microbiol Biotechnol 13:99–103
36. Vollmer AC, Belkin S, Smulski DR, Van Dyk TK, LaRossa RA (1997) Detection of DNA damage by use of *Escherichia coli* carrying *recA'::lux*, *uvrA'::lux*, or *alkA'::lux* reporter plasmids. Appl Environ Microbiol 63:2566–2571
37. Hever N, Belkin S (2006) A dual-color bacterial reporter strain for the detection of toxic and genotoxic effects. Eng Life Sci 6:319–323
38. Mitchell RJ, Gu MB (2004) An *Escherichia coli* biosensor capable of detecting both genotoxic and oxidative damage. Appl Microbiol Biotechnol 64:46–52
39. Mitchell RJ, Gu MB (2004) Construction and characterization of novel dual stress-responsive bacterial biosensors. Biosens Bioelectron 19:977–985
40. Gil GC, Mitchell RJ, Chang ST, Gu MB (2000) A biosensor for the detection of gas toxicity using a recombinant bioluminescent bacterium. Biosens Bioelectron 15:23–30
41. Gu MB, Chang ST (2001) Soil biosensor for the detection of PAH toxicity using an immobilized recombinant bacterium and a biosurfactant. Biosens Bioelectron 16:667–674

42. Gu MB, Gil GC, Kim JH (1999) A two-stage minibioreactor system for continuous toxicity monitoring. Biosens Bioelectron 14:355–361
43. Quershi AA, Bulich AA, Isenberg DL (1998) Microtox toxicity test systems – where they stand today. In: Wells P, Lee K, Blaise C (eds) Microscale testing in aquatic toxicology: advances, techniques, and practice. CRC Press, Boca Raton
44. Van Dyk TK, Majarian WR, Konstantinov KB, Young RM, Dhurjati PS, LaRossa RA (1994) Rapid and sensitive pollutant detection by induction of heat shock gene-bioluminescence gene fusions. Appl Environ Microbiol 60:1414–1420
45. Choi SH, Gu MB (2001) Phenolic toxicity-detection and classification through the use of a recombinant bioluminescent *Escherichia coli*. Environ Toxicol Chem 20:248–255
46. Min J, Pham CH, Gu MB (2003) Specific responses of bacterial cells to dioxins. Environ Toxicol Chem 22:233–238
47. Gu MB, Min J, Kim EJ (2002) Toxicity monitoring and classification of endocrine disrupting chemicals (EDCs) using recombinant bioluminescent bacteria. Chemosphere 46:289–294
48. Nunoshiba T, Nishioka H (1991) 'Rec-lac test' for detecting SOS-inducing activity of environmental genotoxic substance. Mutat Res 254:71–77
49. Kostrzynska M, Leung KT, Lee H, Trevors JT (2002) Green fluorescent protein-based biosensor for detecting SOS-inducing activity of genotoxic compounds. J Microbiol Methods 48:43–51
50. Sagi E, Hever N, Rosen R, Bartolome AJ, Premkumar R, Ulber R et al (2003) Fluorescence and bioluminescence reporter functions in genetically modified bacterial sensor strains. Sens Actuators B Chem 90:2–8
51. Norman A, Hestbjerg Hansen L, Sørensen SJ (2005) Construction of a ColD *cda* promoter-based SOS-green fluorescent protein whole-cell biosensor with higher sensitivity toward genotoxic compounds than constructs based on *recA*, *umuDC*, or *sulA* promoters. Appl Environ Microbiol 71:2338–2346
52. Knight AW (2004) Using yeast to shed light on DNA damaging toxins and irradiation. Analyst 129:866–869
53. Knight AW, Keenan PO, Goddard NJ, Fielden PR, Walmsley RM (2004) A yeast-based cytotoxicity and genotoxicity assay for environmental monitoring using novel portable instrumentation. J Environ Monit 6:71–79
54. Walmsley RM (2005) Genotoxicity screening: the slow march to the future. Expert Opin Drug Metab Toxicol 1:261–268
55. Burlage RS, Sayler GS, Larimer F (1990) Monitoring of naphthalene catabolism by bioluminescence with *nah-lux* transcriptional fusions. J Bacteriol 172:4749–4757
56. Heitzer A, Webb OF, Thonnard JE, Sayler GS (1992) Specific and quantitative assessment of naphthalene and salicylate bioavailability by using a bioluminescent catabolic reporter bacterium. Appl Environ Microbiol 58:1839–1846
57. Köhler S, Belkin S, Schmid RD (2000) Reporter gene bioassays in environmental analysis. Fresenius J Anal Chem 366:769–779
58. Ron EZ (2007) Biosensing environmental pollution. Curr Opin Biotechnol 18:252–256
59. Vollmer AC, Van Dyk TK (2004) Stress responsive bacteria: biosensors as environmental monitors. Adv Microb Physiol 49:131–174
60. Magrisso S, Erel Y, Belkin S (2008) Microbial reporters of metal bioavailability. Microb Biotechnol 1:320–330
61. Fujimoto H, Wakabayashi M, Yamashiro H, Maeda I, Isoda K, Kondoh M et al (2006) Whole-cell arsenite biosensor using photosynthetic bacterium *Rhodovulum sulfidophilum* as an arsenite biosensor. Appl Microbiol Biotechnol 73:332–338
62. Maeda I, Yamashiro H, Yoshioka D, Onodera M, Ueda S, Kawase M et al (2006) Colorimetric dimethyl sulfide sensor using *Rhodovulum sulfidophilum* cells based on intrinsic pigment conversion by CrtA. Appl Microbiol Biotechnol 70:397–402
63. Beck R, Burtscher H (1994) Expression of human placental alkaline phosphatase in *Escherichia coli*. Prot Express Purif 5:192–197

64. Bronstein I, Fortin JJ, Voyta JC, Juo RR, Edwards B, Olesen CE et al (1994) Chemiluminescent reporter gene assays: sensitive detection of the GUS and SEAP gene products. Biotechniques 17(172–174):176–177
65. Shao CY, Howe CJ, Porter AJR, Glover LA (2002) Novel cyanobacterial biosensor for detection of herbicides. Appl Environ Microbiol 68:5026–5033
66. Willardson BM, Wilkins JF, Rand TA, Schupp JM, Hill KK, Keim P, Jackson PJ (1998) Development and testing of a bacterial biosensor for toluene-based environmental contaminants. Appl Environ Microbiol 64:1006–1012
67. Yoon KP, Misra TK, Silver S (1991) Regulation of the *cadA* cadmium resistance determinant of *Staphylococcus aureus* plasmid pI258. J Bacteriol 173:7643–7649
68. Heo J, Thomas KJ, Seong GH, Crooks RM (2003) A microfluidic bioreactor based on hydrogel-entrapped *E. coli*: cell viability, lysis, and intracellular enzyme reactions. Anal Chem 75:22–26
69. Choi JW, Park KW, Lee DB, Lee W, Lee WH (2005) Cell immobilization using self-assembled synthetic oligopeptide and its application to biological toxicity detection using surface plasmon resonance. Biosens Bioelectron 20:2300–2305
70. Umehara S, Wakamoto Y, Inoue I, Yasuda K (2003) On-chip single-cell microcultivation assay for monitoring environmental effects on isolated cells. Biochem Biophys Res Commun 305:534–540
71. Balaban NQ, Merrin J, Chait R, Kowalik L, Leibler S (2004) Bacterial persistence as a phenotypic switch. Science 305:1622–1625
72. Lee JH, Mitchell RJ, Kim BC, Cullen DC, Gu MB (2005) A cell array biosensor for environmental toxicity analysis. Biosens Bioelectron 21:500–507
73. Lee JH, Youn CH, Kim BC, Gu MB (2007) An oxidative stress-specific bacterial cell array chip for toxicity analysis. Biosens Bioelectron 22:2223–2229
74. Ben-Israel O, Ben-Israel H, Ulitzur S (1998) Identification and quantification of toxic chemicals by use of *Escherichia coli* carrying *lux* genes fused to stress promoters. Appl Environ Microbiol 64:4346–4352
75. Parsons AB, Brost RL, Ding H, Li Z, Zhang C, Sheikh B et al (2004) Integration of chemical-genetic and genetic interaction data links bioactive compounds to cellular target pathways. Nat Biotechnol 22:62–69
76. Parsons AB, Lopez A, Givoni IE, Williams DE, Gray CA, Porter J et al (2006) Exploring the mode-of-action of bioactive compounds by chemical-genetic profiling in yeast. Cell 126:611–625
77. Fan J, Ren Y (2006) Statistical analysis of DNA microarray data in cancer research. Clin Cancer Res 12:4469–4473
78. Hughes TR, Marton MJ, Jones AR, Roberts CJ, Stoughton R, Armour CD et al (2000) Functional discovery via a compendium of expression profiles. Cell 102:109–126
79. Benovich E (2003) Whole-cell bacterial biosensor panel for acute toxicity detection in water. MSc thesis, The Environmental Studies Program, Faculty of Natural Sciences, The Hebrew University of Jerusalem, Israel
80. Duda RO, Hart PE (1973) Pattern classification and scene analysis. Wiley, New York
81. Elad T, Benovich E, Magrisso S, Belkin S (2008) Toxicant identification by a luminescent bacterial bioreporter panel: application of pattern classification algorithms. Environ Sci Technol 42:8486–8491
82. Bhattacharya S, Jang J, Yang L, Akin D, Bashir R (2007) BioMEMS and nanotechnology-based approaches for rapid detection of biological entities. J Rapid Methods Autom Microbiol 15:1–32
83. Bolton EK, Sayler GS, Nivens DE, Rochelle JM, Ripp S, Simpson ML (2002) Integrated CMOS photodetectors and signal processing for very low-level chemical sensing with the bioluminescent bioreporter integrated circuit. Sens Actuators B Chem 85:179–185
84. Thorsen T, Maerkl SJ, Quake SR (2002) Microfluidic large-scale integration. Science 298:580–584

85. Yoo SK, Lee JH, Yun SS, Gu MB, Lee JH (2007) Fabrication of a bio-MEMS based cell-chip for toxicity monitoring. Biosens Bioelectron 22:1586–1592
86. Inoue I, Wakamoto Y, Moriguchi H, Okano K, Yasuda K (2001) On-chip culture system for observation of isolated individual cells. Lab Chip 1:50–55
87. Fukuda J, Khademhosseini A, Yeh J, Eng G, Cheng J, Farokhzad OC, Langer R (2006) Micropatterned cell co-cultures using layer-by-layer deposition of extracellular matrix components. Biomaterials 27:1479–1486
88. Ruckenstein E, Li ZF (2005) Surface modification and functionalization through the self-assembled monolayer and graft polymerization. Adv Colloid Interface Sci 113:43–63
89. Thirumalapura NR, Ramachandran A, Morton RJ, Malayer JR (2006) Bacterial cell microarrays for the detection and characterization of antibodies against surface antigens. J Immunol Methods 309:48–54
90. Tani H, Maehana K, Kamidate T (2004) Chip-based bioassay using bacterial sensor strains immobilized in three-dimensional microfluidic network. Anal Chem 76:6693–6697
91. Park S, Wolanin PM, Yuzbashyan EA, Silberzan P, Stock JB, Austin RH (2003) Motion to form a quorum. Science 301:188
92. Biran I, Walt DR (2002) Optical imaging fiber-based single live cell arrays: a high-density cell assay platform. Anal Chem 74:3046–3054
93. Kuang Y, Biran I, Walt DR (2004) Living bacterial cell array for genotoxin monitoring. Anal Chem 76:2902–2909
94. Ingham CJ, Sprenkels A, Bomer J, Molenaar D, van den Berg A, van Hylckama Vlieg JET, de Vos WM (2007) The micro-Petri dish, a million-well growth chip for the culture and high-throughput screening of microorganisms. Proc Natl Acad Sci USA 104:18217–18222
95. Fesenko DO, Nasedkina TV, Prokopenko DV, Mirzabekov AD (2005) Biosensing and monitoring of cell populations using the hydrogel bacterial microchip. Biosens Bioelectron 20:1860–1865
96. Flickinger MC, Schottel JL, Bond DR, Aksan A, Scriven LE (2007) Painting and printing living bacteria: engineering nanoporous biocatalytic coatings to preserve microbial viability and intensify reactivity. Biotechnol Prog 23:2–17
97. Hakkila K, Maksimov M, Karp M, Virta M (2002) Reporter genes *lucFF*, *luxCDABE*, *gfp*, and dsred have different characteristics in whole-cell bacterial sensors. Anal Biochem 301:235–242
98. Ivask A, Green T, Polyak B, Mor A, Kahru A, Virta M, Marks R (2007) Fibre-optic bacterial biosensors and their application for the analysis of bioavailable Hg and As in soils and sediments from Aznalcollar mining area in Spain. Biosens Bioelectron 22:1396–1402
99. Lyngberg OK, Stemke DJ, Schottel JL, Flickinger MC (1999) A single-use luciferase-based mercury biosensor using *Escherichia coli* HB101 immobilized in a latex copolymer film. J Ind Microbiol Biotechnol 23:668–676
100. Xu T, Petridou S, Lee EH, Roth EA, Vyavahare NR, Hickman JJ, Boland T (2004) Construction of high-density bacterial colony arrays and patterns by the ink-jet method. Biotechnol Bioeng 85:29–33
101. Barron JA, Rosen R, Jones-Meehan J, Spargo BJ, Belkin S, Ringeisen BR (2004) Biological laser printing of genetically modified *Escherichia coli* for biosensor applications. Biosens Bioelectron 20:246–252
102. Xu L, Robert L, Ouyang Q, Taddei F, Chen Y, Lindner AB, Baigl D (2007) Microcontact printing of living bacteria arrays with cellular resolution. Nano Lett 7:2068–2072
103. Weibel DB, Lee A, Mayer M, Brady SF, Bruzewicz D, Yang J et al (2005) Bacterial printing press that regenerates its ink: contact-printing bacteria using hydrogel stamps. Langmuir 21:6436–6442
104. Rowan B, Wheeler MA, Crooks RM (2002) Patterning bacteria within hyperbranched polymer film templates. Langmuir 18:9914–9917
105. Bjerketorp J, Håkansson S, Belkin S, Jansson JK (2006) Advances in preservation methods: keeping biosensor microorganisms alive and active. Curr Opin Biotechnol 17:43–49

106. Pedahzur R, Rosen R, Belkin S (2004) Stabilization of recombinant bioluminescent bacteria for biosensor applications. Cell Preserv Technol 2:260–269
107. Stocker J, Balluch D, Gsell M, Harms H, Feliciano J, Daunert S et al (2003) Development of a set of simple bacterial biosensors for quantitative and rapid measurements of arsenite and arsenate in potable water. Environ Sci Technol 37:4743–4750
108. Ulitzur S, Lahav T, Ulitzur N (2002) A novel and sensitive test for rapid determination of water toxicity. Environ Toxicol 17:291–296
109. Fesenko DO, Nasedkina TV, Chudinov AV, Prokopenko DV, Yurasov RA, Zasedatelev AS (2005) Alginate gel biochip for real-time monitoring of intracellular processes in bacterial and yeast cells. Mol Biol 39:84–89
110. Sakaguchi T, Morioka Y, Yamasaki M, Iwanaga J, Beppu K, Maeda H et al (2007) Rapid and onsite BOD sensing system using luminous bacterial cells-immobilized chip. Biosens Bioelectron 22:1345–1350
111. Held M, Schuhmann W, Jahreis K, Schmidt HL (2002) Microbial biosensor array with transport mutants of *Escherichia coli* K12 for the simultaneous determination of mono-and disaccharides. Biosens Bioelectron 17:1089–1094
112. Matsui N, Kaya T, Nagamine K, Yasukawa T, Shiku H, Matsue T (2006) Electrochemical mutagen screening using microbial chip. Biosens Bioelectron 21:1202–1209
113. Peter J, Hutter W, Stollnberger W, Hampel W (1996) Detection of chlorinated and brominated hydrocarbons by an ion sensitive whole cell biosensor. Biosens Bioelectron 11:1215–1219
114. Akselrod GM, Timp W, Mirsaidov U, Zhao Q, Li C, Timp R et al (2006) Laser-guided assembly of heterotypic three-dimensional living cell microarrays. Biophys J 91: 3465–3473
115. Ashkin A, Dziedzic JM (1987) Optical trapping and manipulation of viruses and bacteria. Science 235:1517–1520
116. Ashkin A, Dziedzic JM, Yamane T (1987) Optical trapping and manipulation of single cells using infrared laser beams. Nature 330:769–771
117. Balagaddé FK, You L, Hansen CL, Arnold FH, Quake SR (2005) Long-term monitoring of bacteria undergoing programmed population control in a microchemostats. Science 309:137–140
118. Galvao TC, de Lorenzo V (2006) Transcriptional regulators à la carte: engineering new effector specificities in bacterial regulatory proteins. Curr Opin Biotechnol 17:34–42
119. van der Meer JR, Tropel D, Jaspers M (2004) Illuminating the detection chain of bacterial bioreporters. Environ Microbiol 6:1005–1020
120. Marqués S, Aranda-Olmedo I, Ramos JL (2006) Controlling bacterial physiology for optimal expression of gene reporter constructs. Curr Opin Biotechnol 17:50–56
121. Mitchell RJ, Ahn JM, Gu MB (2005) Comparison of *Photorhabdus luminescens* and *Vibrio fischeri lux* fusions to study gene expression patterns. J Microbiol Biotechnol 15:48–54
122. Neufeld T, Biran D, Popovtzer R, Erez T, Ron EZ, Rishpon J (2006) Genetically engineered *pfabA pfabR* bacteria: an electrochemical whole cell biosensor for detection of water toxicity. Anal Chem 78:4952–4956
123. Wells M, Gosch M, Rigler R, Harms H, Lasser T, van der Meer JR (2005) Ultrasensitive reporter protein detection in genetically engineered bacteria. Anal Chem 77:2683–2689
124. Wells M (2006) Advances in optical detection strategies for reporter signal measurements. Curr Opin Biotechnol 17:28–33
125. Phadtare S, Alsina J, Inouye M (1999) Cold-shock response and cold-shock proteins. Curr Opin Microbiol 2:175–180
126. Storz G, Imlay JA (1999) Oxidative stress. Curr Opin Microbiol 2:188–194
127. Workman CT, Mak HC, McCuine S, Tagne J-B, Agarwal M, Ozier O et al (2006) A systems approach to mapping DNA damage response pathways. Science 312:1054–1059
128. Yura T, Nakahigashi K (1999) Regulation of the heat-shock response. Curr Opin Microbiol 2:153–158

129. Ahn JM, Mitchell RJ, Gu MB (2004) Detection and classification of oxidative damaging stresses using recombinant bioluminescent bacteria harboring *sodA:: pqi:: and katG:: luxCDABE* fusions. Enzyme Microb Technol 35:540–544
130. Galluzzi L, Karp M (2006) Whole cell strategies based on *lux* genes for high throughput applications toward new antimicrobials. Comb Chem High Throughput Screen 9:501–514
131. Biran I, Rissin DM, Ron EZ, Walt DR (2003) Optical imaging fiber-based live bacterial cell array biosensor. Anal Biochem 315:106–113
132. Popovtzer R, Neufeld T, Biran D, Ron EZ, Rishpon J, Shacham-Diamand Y (2005) Novel integrated electrochemical nano-biochip for toxicity detection in water. Nano Lett 5: 1023–1027
133. Reifferscheid G, Heil J (1996) Validation of the SOS/*umu* test using test results of 486 chemicals and comparison with the Ames test and carcinogenicity data. Mutat Res 369: 129–145
134. Lucchini S, Thompson A, Hinton JCD (2001) Microarrays for microbiologists. Microbiology 147:1403–1414
135. Tong AH, Evangelista M, Parsons AB, Xu H, Bader GD, Pagé N et al (2001) Systematic genetic analysis with ordered arrays of yeast deletion mutants. Science 294:2364–2368
136. Winzeler EA, Shoemaker DD, Astromoff A, Liang H, Anderson K, Andre B et al (1999) Functional characterization of the *S. cerevisiae* genome by gene deletion and parallel analysis. Science 285:901–906
137. Tong AH, Lesage G, Bader GD, Ding H, Xu H, Xin X et al (2004) Global mapping of the yeast genetic interaction network. Science 303:808–813
138. Scarcelli JJ, Viggiano S, Hodge CA, Heath CV, Amberg DC, Cole CN (2008) Synthetic genetic array analysis in *Saccharomyces cerevisiae* provides evidence for an interaction between *RAT8/DBP5* and genes encoding P-body components. Genetics 179:1945–1955
139. van Welsem T, Frederiks F, Verzijlbergen KF, Faber AW, Nelson ZW, Egan DA et al (2008) Synthetic lethal screens identify gene silencing processes in yeast and implicate the acetylated amino terminus of Sir3 in recognition of the nucleosome core. Mol Cell Biol 28: 3861–3872
140. Makhnevych T, Sydorskyy Y, Xin X, Srikumar T, Vizeacoumar FJ, Jeram SM et al (2009) Global map of SUMO function revealed by protein-protein interaction and genetic networks. Mol Cell 33:124–135
141. Davierwala AP, Haynes J, Li Z, Brost RL, Robinson MD, Yu L et al (2005) The synthetic genetic interaction spectrum of essential genes. Nat Genet 37:1147–1152
142. Sopko R, Huang D, Preston N, Chua G, Papp B, Kafadar K et al (2006) Mapping pathways and phenotypes by systematic gene overexpression. Mol Cell 21:319–330
143. Chang M, Bellaoui M, Boone C, Brown GW (2002) A genome-wide screen for methyl methanesulfonate-sensitive mutants reveals genes required for S phase progression in the presence of DNA damage. Proc Natl Acad Sci USA 99:16934–16939
144. Xie MW, Jin F, Hwang H, Hwang S, Anand V, Duncan MC, Huang J (2005) Insights into TOR function and rapamycin response: chemical genomic profiling by using a high-density cell array method. Proc Natl Acad Sci USA 102:7215–7220
145. Typas A, Nichols RJ, Siegele DA, Shales M, Collins SR, Lim B et al (2008) High-throughput, quantitative analyses of genetic interactions in *E. coli*. Nat Methods 5:781–787
146. Watanabe K, Joh T, Seno K, Sasaki M, Todoroki I, Miyashita M et al (2001) Development and clinical application of an immunoassay using intact *Helicobacter pylori* attached to a solid phase as an antigen. Clin Biochem 34:291–295
147. Ison CA, Hadfield SG, Glynn AA (1981) Enzyme-linked immunosorbent assay (ELISA) to detect antibodies in gonorrhoea using whole cells. J Clin Pathol 34:1040–1043
148. Oh SH, Lee SH, Kenrick SA, Daugherty PS, Soh HT (2006) Microfluidic protein detection through genetically engineered bacterial cells. J Proteome Res 5:3433–3437

Adv Biochem Engin/Biotechnol (2010) 117: 109–130
DOI:10.1007/10_2009_2

Published online: 21 May 2009

Surface Functionalization for Protein and Cell Patterning

Pascal Colpo, Ana Ruiz, Laura Ceriotti, and François Rossi

Abstract The interaction of biological systems with synthetic material surfaces is an important issue for many biological applications such as implanted devices, tissue engineering, cell-based sensors and assays, and more generally biologic studies performed ex vivo. To ensure reliable outcomes, the main challenge resides in the ability to design and develop surfaces or artificial micro-environment that mimic 'natural environment' in interacting with biomolecules and cells without altering their function and phenotype. At this effect, microfabrication, surface chemistry and material science play a pivotal role in the design of advanced *in-vitro* systems for cell culture applications. In this chapter, we discuss and describe different techniques enabling the control of cell-surface interactions, including the description of some techniques for immobilization of ligands for controlling cell-surface interactions and some methodologies for the creation of well confined cell rich areas.

Keywords Surface chemistry • patterning techniques • microfabrication • self-assembled monolayers • plasma polymers • soft lithography

Contents

P. Colpo (✉), A. Ruiz, L. Ceriotti, and F. Rossi
European Commission, Joint Research Centre, Institute for Health and Consumer Protection, Via E.Fermi, 2749 TP203, 21027 Ispra, Varese, Italy
e-mail: pascal.colpo@jrc.it

1 Introduction

The interaction of biological systems with synthetic material surfaces is an important issue for many biological applications such as implanted devices, tissue engineering, cell-based sensors and assays, and more generally biologic studies performed ex vivo [1–4].

In vitro assays are based on culturing cells on artificial surfaces that have properties very different from the cell natural environment, which makes assay outcomes uncertain due to the high complexity of all phenomena involved [5]. Therefore, success of in vitro assays reside in the ability to develop surfaces or artificial microenvironments that mimic the "natural environment" interacting with biomolecules and cells (cell lines or primary culture cells) in a specific mode without altering their function and phenotype and that could control and drive the cell response on surfaces.

Many examples of the use of microfabricated platforms for in vitro assays can be found in the literature. Platforms with advanced surface chemistries have been successfully developed for instance to drive differentiation of stem cells [6, 7], for studying the factors that affect cell apoptosis [8] and cell structural organization and division [9, 10]. Micropatterned cocultures have been shown to be valuable platforms for long term hepatocyte culture and for the miniaturization of hepatoxicity screening assays [11]. Eukaryotic and prokaryotic cells on solid surfaces are also widely used in cell based sensor devices to detect foreign compounds such as toxic and pathogenic agents ([12, 13], and Chapter by Yosi Shacham in this book).

These examples show that microfabrication, surface chemistry and material science play a pivotal role in the design of in vitro systems for cell culture applications.

In this chapter, we discuss and describe different techniques used to functionalize and pattern surfaces enabling the control of the interactions between cells and substrates. The first part deals with the description of the techniques for immobilization of ligands for controlling cell-surface interactions. The second part concerns some methodologies used to pattern surfaces for the creation of well confined cell rich areas.

2 Surface Chemistry to Control Immobilization of Ligands on Surfaces

The interactions between eukaryotic cells and engineered surfaces are mediated by integrins, which are the major transmembrane proteins that link the cell cytoskeleton to specific ligands present on the surface (i.e., proteins of the extra cellular matrix (ECM)). The development of surfaces enabling the control of the ligand types, density, spatial distribution and conformation is therefore crucial to control cell behavior on surfaces [14, 15]. In particular, an important issue is to control the overall ligand capacity to promote specific integrin binding while limiting the nonspecific adsorption [16]. Such surface activity depends very much on the surface chemistry of the support. For instance, ECM proteins adsorbed on treated or untreated tissue culture polystyrene substrates (TCPS) may have different bioactivities due to the differences of surface wettability [17]. Indeed, proteins may undergo conformation change, denaturation and have different orientation upon adsorption, leading to poor control of the overall surface bioactivity [18, 135]. Furthermore, initially immobilized proteins can be displaced and removed from the surface by proteins from culture media or proteins produced by cells during the ECM remodeling process. This means that, depending on the molecular weight and bioaffinity of the proteins present in the medium, the protein layer identity can be modified during the course of the experiments by conformational changes and ligand substitution [19]. This underlines the need for surfaces with stable and exposed ligands for specific interaction with cells, together with areas where limited nonspecific adsorption. The following subsections will describe two techniques enabling the fabrication of adhesive and anti adhesive layers, namely self assembled monolayers and plasma polymerization.

2.1 *Self Assembled Monolayers*

Self-assembled monolayers (SAMs) are the most used vehicle to produce model surfaces to study protein- and cell-surface interactions. SAMs are based on molecules that have the property to self–assemble on solid surfaces forming a monolayer with a well-defined chemistry. There are two main classes of SAMs: alkanethiols (HS(CH2)*n*X) that are normally deposited on gold and silver [20] and the alkylsilanes and alkylsiloxanes deposited on hydroxylated surfaces [21] (Fig. 1). SAMs head groups can have a vast variety of chemical functions such as –COOH, –OH, $–NH_2$, $–CH_3$, etc. providing a versatile tool to study interactions between cells, proteins and surfaces. Nevertheless, the formation of a high ordered SAM without defects is a difficult task and highly related to the purity of the solutions as well as the "quality of the substrate", i.e., cleanliness, roughness and crystallinity [20].

Many examples can be found in the literature on the use SAMs as platforms to control adhesion migration and differentiation of cells.

2.1.1 Influence of Surface Chemistry on Cell Response

Extensive work has been performed by the Garcia group on the influence of functionalized surfaces on fibronectin conformation and subsequent cell adhesion,

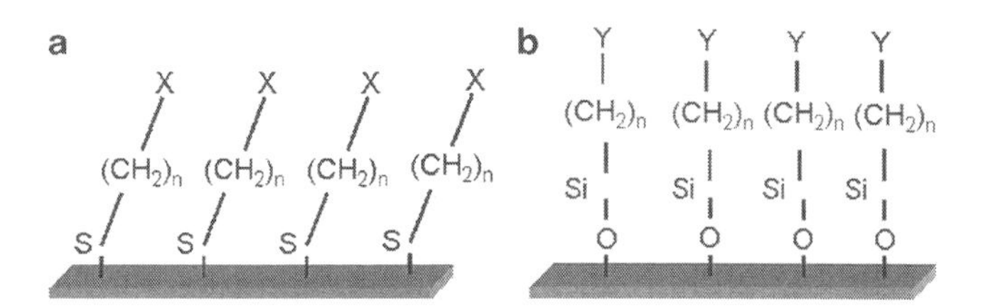

Fig. 1 A schematic representation of SAMs organothiols on gold (**a**) and organosilane on SiO_2 surface (**b**) where X and Y denote the chemical functionality of the final surface

proliferation and differentiation [14]. In these works, fibronectin was adsorbed on gold coated with alkanethiol SAMs with different headgroups (CH_3, OH, COOH, NH_2). Results showed that the fibronectin structure was particularly sensitive to the surface chemistry, resulting in different functional activities [22]. For instance, the authors observed a selective integrin binding to $\alpha5\beta1$ integrin for OH and NH_2 surfaces, binding to both $\alpha5\beta1$ and $\alpha V\beta3$ for the COOH surfaces, and poor binding on the CH_3 functionalized supports. Functionalized surface-mediated fibronectin adhesion enables the modulation of osteoblastic differentiation and mineralization [23] and also regulation of the switch between myoblast proliferation and differentiation [24, 25] (Fig. 2).

Carboxylic acid (COOH) and amine (NH_2)-terminated SAMs of alkanethiolates have been used to study the influence of the surface charge on the modulation of cell adhesion and spreading through the control of the orientation of adsorbed $FnIII_{7–10}$, a cell-adhesive protein containing RGD (R: arginine; G: glycine; D: aspartic acid) residues which are commonly recognized by the integrins.

Results indicate that NH_2–SAMs orientate the $FnIII_{7-10}$ in such a way that more bovine aortic endothelial cells adhere and spread on the NH_2–SAM as compared to COOH–SAMs coated surfaces [26]. Besides, the control of the orientation of osteopontin on OH, COOH, NH_2 and CH_3 terminated SAMs has been performed in order to modulate endothelial cell adhesion [27]. Self-assembled monolayers of alkylsilanes with CH_3, polyethylene glycol (PEG), COOH, NH_2 and OH terminations have been used to study human fibroblast adhesion [28]. Better attachment, spreading and growth of human fibroblasts on COOH and NH_2 SAMs were found as compared to the other SAMs. The better interaction of fibroblasts was related to a better integrin activity as observed after antibody-tagging of living cells.

These examples shows how cell response can be influenced and modulated by ECM protein conformation through surface chemistry.

An alternative strategy consist of using RGD peptides directly immobilized on the surface. RGD is a cell adhesion promoter which presents many advantages in term of simplicity and stability as compared to entire ECM proteins [136]. Numerous materials have been functionalized with RGD peptides. An outstanding review on the used of RGD modified polymer to functionalize surface can be found elsewhere [29].

2.1.2 Control of Ligand Density and Distribution

A problem in studying the effects of immobilized ligands on cell–surface interaction is that adherent cells tend to readapt their microenvironment by remodeling the extracellular matrix in long term cultures modifying the ligand nature during the experiments. To address this issue, a strategy has been developed by generating surfaces containing two different types of SAMs, one allowing the ligand-directed binding of cells and the other resisting the protein and cell adhesion (Fig. 3) [30]. In this work, the surfaces were made from alkanethiolates containing RGD peptides as ligands to promote integrin-mediated cell adhesion, and oligo(ethyleneglycol) moieties that prevent the nonspecific

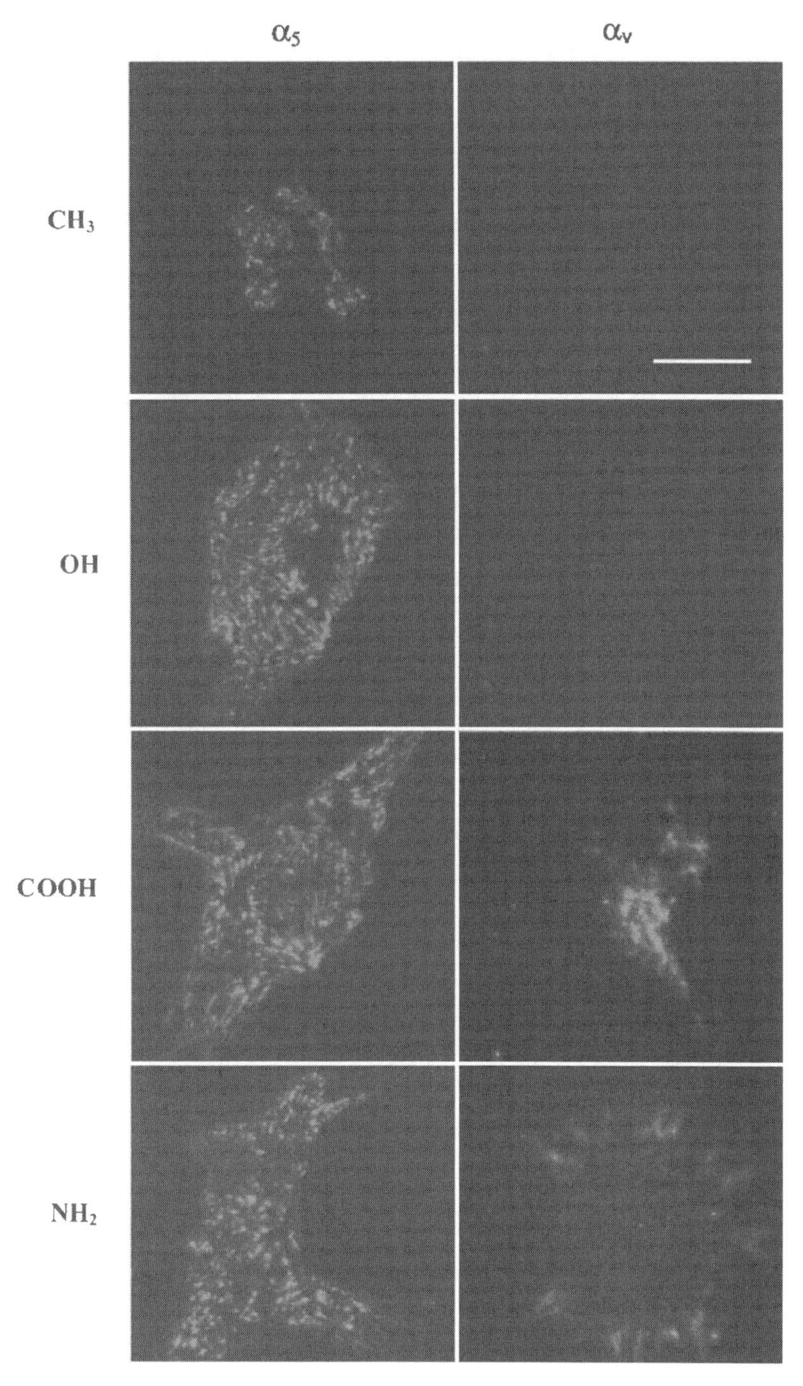

Fig. 2 Influence of surface chemistry on myoblast integrin binding to FN coated SAMs (from [24, 25], Elsevier)

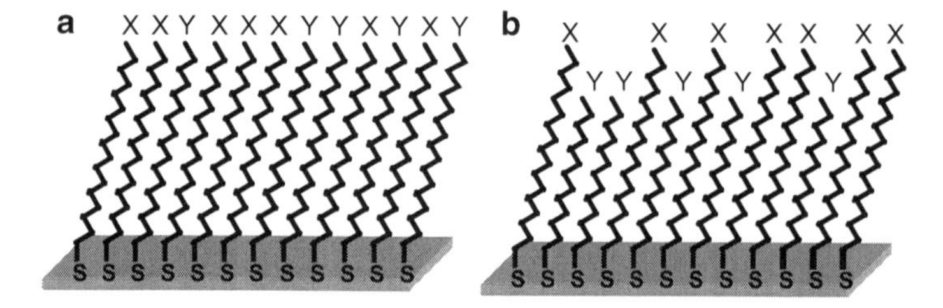

Fig. 3 Representation of self-assembled monolayer of alkanethiolates with two different terminal groups (X,Y). The nature of head groups presented at the surface can be based on oligo(ethyleneglycol) to reduce nonspecific adsorption (Y), and bioactive (COOH, NH2) or RGD containing head groups (**a**). Chains length of oligo(ethyleneglycol) can be varied to increase ligand accessibility (**b**)

adsorption of proteins and cells. The authors showed that the number of bovine capillary endothelial cells spreading on mixed SAM surfaces reaches a plateau at SAMs-RGD ratio above 0.001 with a reduced deposition of proteins by attached cells. They show that ? (low amount of) RGD presenting surfaces in a inert background is sufficient to promote long-term attachment and survival of cells.

Houseman and Mrksich [31] used a surface composed of SAMs presenting RGD in a background of oligo(ethylene glycol) SAMs with different chain lengths (Fig. 3b). They showed that the microenvironment, i.e., the binding site accessibility of the immobilized ligands, influences the attachment and spread of Swiss 3T3 fibroblasts. This study clearly shows that the configuration of adhesive motifs can be precisely designed and can influence the response of mammalian cells.

The spatial distribution of RGD peptides can be controlled as well by using surface nanopatterning [32]. The method is based on the deposition of gold containing diblock copolymer micelles on glass surfaces, followed by hydrogen gas plasma treatment to remove the polymer that leave gold nanoparticles of 5–8 nm in diameter on the surface. The remaining region is passivated by using PEG silanes. The gold nanodots are subsequently functionalized via an alkane thiol presenting a c-RGDfK peptide sequence present in many of ECM protein and responsible of the integrin binding. Cell cultures were performed with rat embryonic fibroblasts on this biofunctionalized gold template with spacing of 58 and 110 nm. The results show that rat embryonic fibroblasts spread very well on the 58–nm pattern with the expected morphology whereas limited cell spreading is observed on the 110–nm pattern.

2.1.3 SAMs Based Antiadhesive Surfaces

Surfaces that resist to non specific adsorption of proteins are fundamental for applications requiring long term cell culture [30] and long term cell confinement in well defined areas [33, 34].

The most widely used bioresistant surfaces are based on polyethylene glycol moieties (PEG) (also known as poly(ethylene oxide) or PEO) having [–CH2–CH2–O–] as monomeric repeat unit.

PEG molecules are usually coupled to specific end-groups or combined with other polymers for long term and stable immobilization on surfaces since PEG chains alone tend to adsorb only weakly on most surfaces. Oligo-PEG and PEG grafted onto alkanethiolate SAMs are the most commonly used surface functionalizations [35, 36] and have been used successfully as protein resistant surfaces. Chain density, length and conformation of the films have been identified as important parameters that make PEG antiadhesive [37]. However, the mechanisms of the antiadhesive properties is not yet well understood [38–41].

Polycationic poly(l–lysine)-*g*-poly(ethylene glycol) (PLL*g*–PEG) has strong anti adhesive properties and is an interesting alternative since it adsorbs electrostatically to negatively charged surfaces (metal, glass, oxides, tissue culture treated polystyrene), a common case of a wide range of biology supports [42].

Extensive work has been done by the group of Whitesides and coworkers on screening a large set of functional groups other than ethylene glycol that resist adsorption of proteins and bacteria [43–45]. Best antiadhesive properties were found in surfaces that were hydrophilic, containing groups that were hydrogen-bond acceptors and with an overall electrically neutrality. The same group surveyed SAMs and polymeric films that is resistant to bacteria, mammalian cells and proteins. Antiadhesive properties of protein resistant SAMs terminated with different functional groups were tested with bacteria (*Staphylococcus aureus, Staphylococcus epidermidis*) and mammalian cells (bovine capillary endothelial (BCE) cells) [44, 45]. Antiadhesive properties of tri(sarcosine) and *N*-acetylpiperazine to bacteria were similar to those obtained with $(EG)_3OH$ and only the tri(sarcosine) terminated SAMs resist the adhesion of BCE cells. Thin polymeric films based on polyamines (poly(ethylenimine)) were grafted and converted to an inert form by acylation, producing films that resist the adsorption of protein and the adhesion of bacteria [46]. Better protein resistance was found by functionalizing the polyamine with acyl chlorides that were derivatives of oligo(ethylene glycol). These polymers have protein and bacteria resistant properties that are comparable or better than $(OG)_3OH$ SAMs.

The duration of the antiadhesive characters of SAMs oligo(EG)-terminated thiols is an issue when long term cell culture is required. Desorption of the film and degradation of its properties can occur and be accelerated by cell processes [30].

The stability of different undecanethiol and tri(ethylene glycol)-terminated undecanethiol) SAMs was studied in phosphate-buffered saline and calf serum [47]. The samples were characterized over 35 days by using contact angle measurements and cyclic voltammetry. Voltammetric measurements changed dramatically for samples in both PBS and calf serum after 21 days as a results of loss of integrity of the SAMs surface coverage. This stability in time could be extended up to 30 days by using palladium as substrate [48].

Micropatterned surfaces have been used to study the role of cellular processes on the degradation of the patterns [49]. Different nonadhesive layers such as agarose, pluronics, hexa(ethylene glycol), or polyacrylamide have been used as platforms to culture cells in serum. Polyacrylamide remained inert and cells patterns remained intact for at least 28 days whereas agarose and pluronics lose their nonadhesive properties due to film desorption, independently of the presence of cells.

2.2 Plasma Polymerization

Plasma polymerization is an interesting alternative for surface functionalization [50–54]. With this technique, conformal thin polymeric films can be deposited on materials with a wide range of chemical functions that can promote adhesion, spreading and proliferation of cells or alternatively that can repel proteins and cells from surfaces.

Plasma polymers are deposited using a glow discharge created from monomer vapor generally by using capacitively coupled plasma sources. The glow discharge containing ions, free electrons and monomer fragments is generated by a high frequency electric field (generally 13.56 MHz) applied between two parallel plate electrodes (Fig. 4). Properties of the films can be controlled by tuning the discharge parameters, particularly pressure, gas residence time and power [52–54].

Many functional films can be produced by plasma polymerization [55], e.g., plasma-deposited acrylic acid (pdAA) with carboxylic acid function [56, 57, 58, 59] allylamine [60, 61], aldehyde [62] and epoxy functionalized surfaces [63] whereas protein and cell resistant surfaces can be fabricated by plasma-deposited poly(ethylene oxide)-like (PEO-like) materials [52–54, 64].

For instance, pdAA films with controlled surface density of carboxyl groups induce attachment and growth of keratinocytes, osteoblasts and fibroblasts [58, 65].

Poly(ethylene oxide) (PEO)-like coatings can be plasma deposited from precursors having $(CH_2–CH_2–O)n$ monomeric repeat units. Plasma deposited PEO-like films have antiadhesive properties at low monomer fragmentation, i.e., low input power and high monomer flow [66]. The fragmentation degree is characterized by X-ray photoelectron spectroscopy (XPS) where the C1s peak of the polymer is compared to the monomer value [67]. It is shown that the C–O–C peak intensity relates directly to the antifouling properties of the film. Adsorption and retention of proteins on plasma deposited poly(ethylene glycol)-like films from different monomers (mono–di–tri–tetra-glymes, and cyclic precursors dioxane, and crown

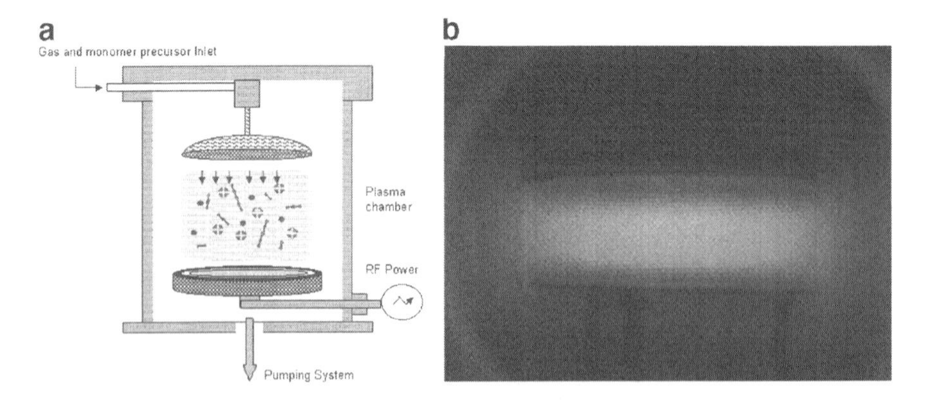

Fig. 4 Schematic representation of Plasma reactor (**a**) and picture of an argon plasma (**b**)

ethers) were measured by radio labeling techniques. Whereas PEO-like films deposited with monoglyme and dioxane adsorbed a large amount of proteins (respectively 71 and 87 ng cm^{-2}), films deposited with others precursor adsorbed low level of proteins (from 1.6 to 15 ng cm^{-2} m^{-2}). Better results were found with triglymes and tetra-glymes with proteins adsorption lower than 3 ng cm^{-2}.

An interesting property of PEO-like films is that its antiadhesive properties can be modulated by tuning the parameters of the plasma during the PEO-like film deposition [52–54] or by modifying its antiadhesive properties by plasma treatments [68]. It can also be combined with adhesive layers by using lithography techniques, e.g., photolithography for cell patterning [52–54].

3 Surface Patterning

Surface patterning is of great importance in many fields of science and technology. Originally developed by the microelectronics industry, it is now being applied to many other different fields such as optics and biology. Generation of patterns is usually accomplished by serial techniques that are able to provide arbitrary features with different physico-chemical properties on the surface of a substrate. Replication of patterns is done by transferring the structural information of a mask or master into another material with the possibility of obtaining many copies of the original structure [137]

Micro- and nanopatterning techniques have been extensively used for biological applications, to engineer surfaces with chemical contrast that allow the confinement of biological materials in predetermined positions. Patterning consists of creating domains that readily adsorb proteins, and hence are cell-adhesive, in a background that is nonfouling and cell-repellent. For instance, micropatterned surfaces are widely used as cell culture platform to study the influence of cell microenvironment on cell response [69]. Seminal work on cell micro patterning has been performed by Chen et al. [8] wherein influence of patterned size on cell response (apoptosis, growth and proliferation) was studied. More recently, the design of patterns with different sizes and shapes allowed to study cell division and axis orientation [10] and to direct cell migration on asymmetric patterns [48]. When high throughput analysis is the objective, i.e., in cytotoxicity and drug screening, cell microarrays is the technique of choice to replace multiwell plates [33].

Production of biopatterns is normally made by conventional photolithography, soft lithography and microarray technologies, and at the nanoscale by scanning probe lithography techniques and nanosphere lithography [70]. Micro- and nanopatterned substrates are being exploited in biosensing for creating cell arrays for high-throughput screening, and to study the roles of focal adhesions formations and cell–cell interactions on different cell developmental processes [32, 71]. One of the key problems of cell patterning is the stability of these patterns, since cell secretions alter the cell adhesive layer.

This paragraph describes the most commonly used methods to micropattern surfaces for bacteria and mammalian cell cultures. Many methodologies for surface patterning are being used; among the most popular are photolithography and soft lithography [69, 72]. Ablation of antiadhesive polymers [73], photocleavage of protecting groups [74] or functionalization of cell repellent polymeric materials by plasma [68, 75] and UV light [138, 76] are also methods used to create patterned surfaces for promoting cell adhesion.

3.1 Photolithography

Photolithography involves the exposure of a photosensitive material through a mask with the pattern of interest. In the case of positive photoresist, the UV light exposed regions become soluble and are subsequently developed. In a second step, the deposition or etching of materials of interest can be performed, followed by the photoresist lift off with the adapted solvents. The resolution (size of the transferred features) depends on the light wavelength used for the illumination. This patterning technique was first applied in biology for the fabrication of DNA arrays [77, 78]. This method is well adapted to fabricate surfaces with two different materials having different properties for selective surface functionalization or biomolecules immobilization. For instance, gold microislands on silicon substrate allow the selective functionalization of the two materials with bioadhesive thiolated SAMs and antiadhesive PEG silanes [24, 25] for cell micropatterning. Surface patterning of deformable, solvated substrates has also been achieved by photolithography [79, 80]. In these works, a photolithographic method was used to obtain PEG hydrogels patterns in which the acrylated moieties in the precursor are conjugated by UV exposure through a transparency mask, giving the possibility of multiple peptide immobilization on the same surface. The same group developed a method for the fabrication of 3D cross-linked PEG hydrogel patterns to create high density microscale wells containing one to three hepatocyte cells in each well [81].

Photolithography can advantageously be combined with plasma polymerization to fabricate surface with microscale chemical contrast [52–54, 82, 83], leading to a selective cell immobilization (Fig. 5).

The main disadvantages of photolithography are the high cost of equipment and the access to a clean room. Moreover, this technique cannot be used to pattern directly proteins and cells on the surfaces, and cannot be applied to nonplanar substrates.

3.2 Soft Lithography

Soft lithography groups together a set of techniques based on molding and printing by using soft materials. The use of soft lithography for protein and cell patterning has been developed and studied in depth by the group of Whitesides [84, 85]. Soft lithography techniques, such as microcontact printing, micromolding in capillaries, patterning using microfluidic channels and laminar flow patterning have been used

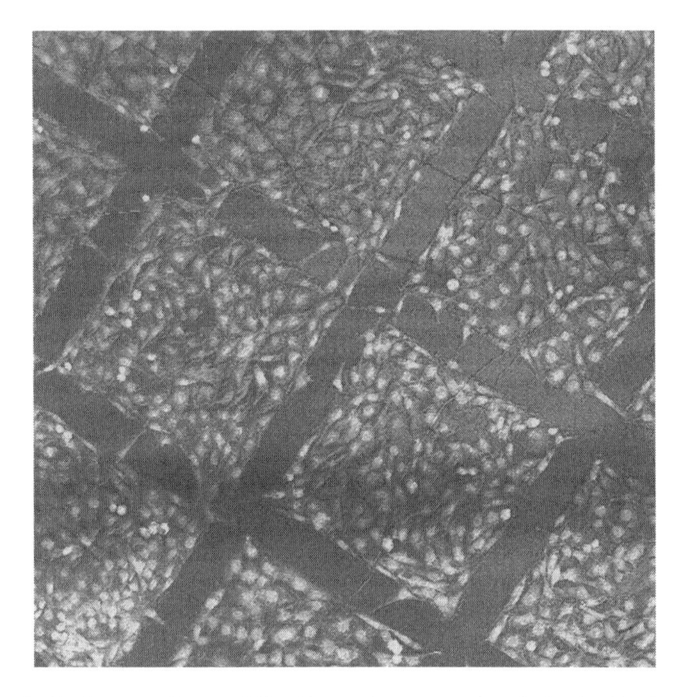

Fig. 5 Confocal microscope image of fibroblasts micropatterned on fouling/antifouling contrasted surfaces produced by plasma polymerization of polyethylene oxide like film and photolithography

successfully to control surface chemistry and cell environment. In replica molding [86–88] a polymer is casted against a patterned master and the 3D topography of the rigid master is replicated in the elastomer with a resolution down to tenths of a nanometer. Elastomeric stamps and molds are usually fabricated in polydimethylsiloxane (PDMS), which has the important properties of having a moderate stiffness, being nontoxic, biocompatible, optically transparent and available commercially at low cost. Furthermore, its surface wettability can be tuned from hydrophobic to hydrophilic by short plasma treatments, giving more flexibility for printing efficiency improvement.

In microcontact printing, the most representative soft lithography technique, thc PDMS stamps are inked with the printing material (SAMs or biomolecules) and put in conformal contact with a substrate. The material on the bas-reliefs of the PDMS that is in contact with the substrate is transferred to the substrate. The PDMS technology allows the obtaining of high quality patterns routinely with feature sizes in the micron level. This so-called microcontact printing method (Fig. 6) was first described by Whitesides and coworkers and used for patterning SAMs on gold surfaces [89]. Then, the technique was further developed to print monolayers of proteins directly on surfaces and the bioactivity of the printed proteins was demonstrated [90–93].

The spatial distribution of adhesion molecules and extracellular matrix proteins on surfaces influences cell behavior. A number of papers have reported on the advantages of the use of such micropatterned environments for cell biology studies

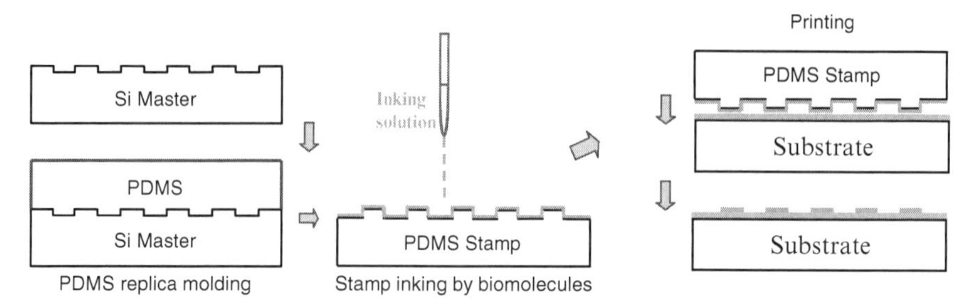

Fig. 6 Schematic drawing of the microcontact printing process

[69, 94, 95]. Micropatterned substrates fabricated by microcontact printing have been used successfully to guide cell migration of fibroblasts [96], to control growth, apoptosis and differentiation of capillary endothelial cells [8, 97] and to study spreading of melanoma mouse cells [98].

Moreover, microcontact printing can be used for direct patterning of biomolecules on antifouling surfaces [99–104]. Indeed, PEO-like films deposited by plasma polymerization resist protein adsorption in aqueous solution but accept proteins printing in dry conditions. By means of this surface biofunctionalization, contrasted fouling/antifouling areas are easily created, which offers improved capabilities for the confinement, localization and guided growth of cells, as well as for studying different cell developmental processes in stem cells [34]. For instance, micropatterned surfaces for single cell study (human umbilical cord blood-derived neural stem cells HUCB-NSC) are being performed in our laboratory by using 10-μm polylysine features microcontact printed on PEO films deposited by plasma polymerization [99, 100] (Fig. 7).

Similarly, microspotting techniques can be used to fabricate direct bioadhesive/biorepellent surfaces using plasma polymerized PEO surfaces. Varied ECM proteins can also be directly microspotted on PEO-like films to form a microarrayed platform for stem cell studies [139] (Fig. 8).

Other soft lithography approaches of interest for patterning surfaces are based on the use of elastomeric membranes with holes that allow the incubation of biological material and, after lift off of the membrane, the biomolecules are delivered to the substrate through the holes [105]. Soft lithography is also used to fabricate microfluidic channels, in which fluids flow laminarly, which can be used for patterning surfaces as well as for delivering substances locally at the interface between different fluid streams [102–104, 106]. Three-dimensional microfluidic structures achieved by stacking membranes have been used for patterning different materials in one step [107, 108].

Soft lithography is well suited to pattern the composition, topography and properties of surfaces. As compared to conventional photolithography, soft lithography related techniques have the advantages of patterning the relevant biomolecules directly on surfaces, controlling the molecular structure of surfaces and the possibility of fabricating microchannels for microfluidics. Moreover, it is inexpensive, rapid, convenient and easy to use.

Fig. 7 Fluorescence microscopy image of 10 × 10 μm square PLL patterns (**a**) and optical microscopy image of HUCB-NSC after 1 day culturing (**b**)

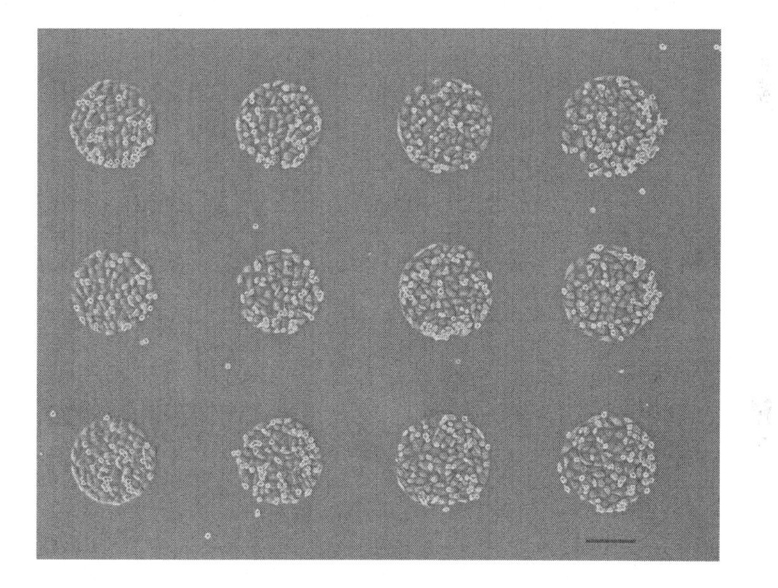

Fig. 8 HUCB-NSC microarray on four and five drops-spots of fibronectin spotted on PEO like film. The *scale bar* is 200 μm

Some new methodologies have recently been developed to pattern two or more cell types in spatially defined cocultures [109, 110] in order to study cell-matrix and cell-cell interactions. Very advance chemistry is involved with electroactive polymers [111, 112] or thermally responsive polymers [113, 114] that can alternatively switch from inert to cell adhesive surfaces. Another interesting approach has been developed by using a layer by layer deposition method [110]. In this method, a first layer of hyaluronic acid (HA) is deposited onto glass substrates by spin coating and micropatterned with capillary force lithography [115, 116]. Capillary force lithography is based on the use of a PDMS stamp placed on top of the HA layer. The HA layer located under the void space recedes until the glass become uncovered, creating

HA micropatterning. Then fibronectin is deposited selectively in the bare glass region thanks to the antiadhesive properties of the HA. The first type of cell is plated and selectively adheres to the fibronectin coated regions. Then the HA-covered areas are complexed with collagen, a cationic polymer which makes this region cell adherent. A second type of cell is added to adhere in the HA-collagen covered regions. The advantage of this technique is that it uses only extracellular matrix components, guaranteeing high biological affinity and no cytotoxic effects.

4 Patterning Bacteria

The development of sensing platforms based on bacteria has different requirements in term of immobilization. For bacteria-based biosensors, the main requirements consist in the careful control of cell positioning together with the capability to keep cell viability and activity for long time periods [13, 117]. Initial interaction mechanism of adhesion of bacteria with surfaces is governed by van der Waals and electrostatic interactions. Irreversible adsorption is mediated by interaction of protein layers (lipopolysaccharides) of the bacteria membrane with the substrate [118]. The development of protein resistant surface is therefore the key objective to design surfaces where bacteria are not going to adhere.

A number of methods have been used to create adhesive cues on solid surfaces to pattern bacteria. Most of them are based on soft lithography approaches.

Production of high resolution arrays of living bacteria has been facilitated by microcontact printing with high aspect ratio PDMS stamps, by selectively transferring bacteria from an agarose substrate to another support. With this method, single bacteria arrays were produced over large areas [119]. Microcontact printing has also been used to fabricate biomimetic micropatterned surfaces of three bacterial proteins (S-layers). First, a nonfouling poly-l-lysine grafted PEG layer is microcontact printed on the surface followed by a back filling with S-layer proteins to which bacteria attach [120]. Contact printing has been also used by Weiber et al. [121] to print bacteria directly onto agar surfaces. Agarose stamps have been inked with bacteria colonies that are partially transferred to agar plates upon contact. The bacteria remaining on the stamp are able to regenerate, allowing the reuse of the agarose stamps for several months.

Alternative methods to microcontact printing have also been reported for patterning bacteria. Using a micromolding in capillaries (MIMIC)-based methods, a polyelectrolyte surface (PEL) can be patterned by micromolding PEG in capillaries. PEG is flown inside microfluidic channels created on the PEL surface and photopolymerized. The PEL/PEG contrasted regions improve the fluorescence intensity of proteins attached to the PEL areas. This simple preparation of functionalized surface has be applied to various biomolecules such as proteins, bacteria, and cells. [122]. Utilizing host–parasite and virus–antibody interactions, Suh et al. [115, 116] were able to separate infected bacteria. By capillary lithography, they produced wells filled with an antibody against virus proteins. Infected bacteria selectively adhered to the patterns of the antibody by specific recognition between the antibody and the virus present in the infected bacteria membrane. Antibody–antigen interactions have also been used for patterning bacteria by microcontact printing of antibodies on silanized glass [123] or by patterning antibodies using Ga^+ ion etching [124]. In the latter, micropatterns of antibodies have been produced by etching a nonfouling layer using a programmable focused Ga^+ ion beam. The etched regions are filled with a cross-linker that links an antibody, the anti-CFA/I, which binds to CFA/I, an antigen protruding outside the bacteria cell membrane.

As mentioned previously, plasma polymers can be advantageously combined with photolithography to create well define micropatterns. For instance, micropatterned surfaces combining a bacteria adhesive region (plasma polymerized allylamine, adhesive polyethylene oxide (PEO*)) and an antiadhesive region (plasma polymerized polyethylene oxide) have been tested with HB101 and TA15 bacteria strains leading to well defined bacteria micropatterns (Fig. 9).

Bacteria were also patterned using a colloidal lithography approach [125]. PDMS templates of 2D arrays of different structures have been fabricated embed-

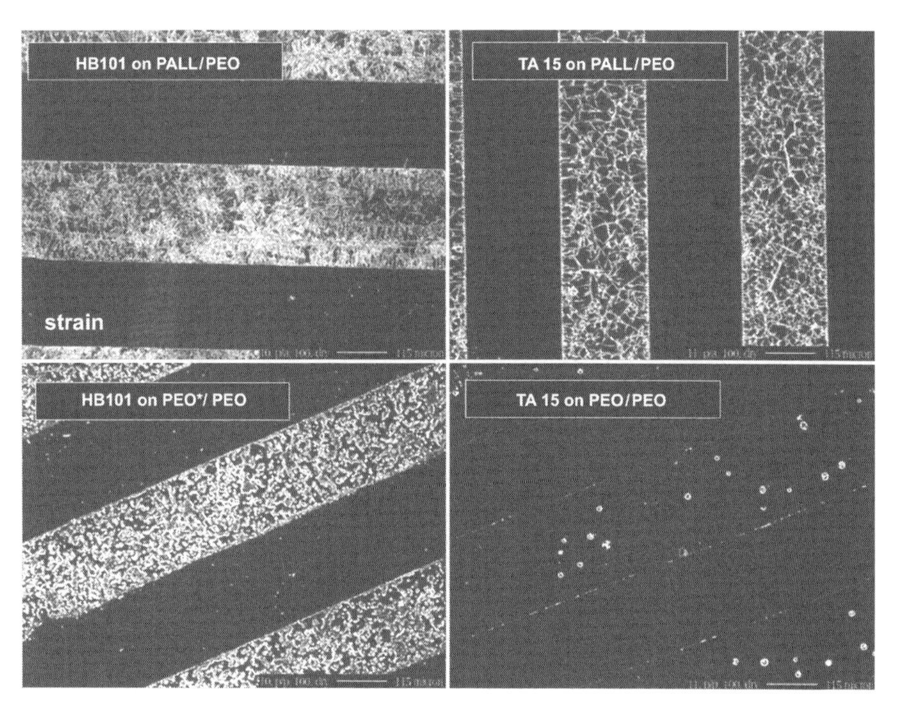

Fig. 9 HB101 and TA15 bacteria strains patterned of plasma polymers (PEO*, adhesive PEO like films)

ding nanoparticles in PDMS. After etching of the nanoparticles, surfaces of PDMS/glass contrast were obtained. Bacteria attachment to the PDMS regions is attributed to hydrophobic interactions.

On the other hand, three-dimensional networks of different types of living cells have been assembled in hydrogel using arrays of time-multiplexed, holographic optical traps. The array of optical tweezers was able to pattern arrays of single bacteria with demonstrated viability [126].

5 Conclusions

Many fabrication methods are already available to design advanced surfaces, enabling the study of cell adhesion, migration and differentiation. Advanced surface chemistries have proven capability to control ligand–integrin interactions allowing the selection between proliferation and differentiation and promoting long term culture. New emerging nanotechnology tools will be key for controlling the number of fibronectin immobilized in each adhesion site, thus controlling the number of integrins involved in each cell–substrate interaction [32, 71]. Single cell or cell monolayer patterns and cell coculture can be obtained by several techniques facilitating the study fundamental aspects of cell–surface and cell–cell interaction as well as improving the reliability and throughput of in vitro assays. In this respect, cell microarrays are being widely used as an alternative to multiwell plates for studying cell adhesion on a large variety of materials [127] or for high throughput analysis of cellular functions and drug screening [6, 33, 128].

This review has focused mainly on functionalization and patterning of flat, rigid two dimensional (2D) surfaces for in vitro assay and cell based sensor application. Nevertheless, these models are known to be insufficient to study and understand complex cell models (i.e., hepatotocytes) which need a "natural-like 3D microenvironment" for their growth and differentiation. The development of a 3D microenvironment matrix with controlled stiffness, nanoscale control of ECM proteins distribution and density can have a major influence on cell response and consequently are becoming in great demand for developing new generations of in vitro assays [129, 130]. The merging of nano- and microtechnology to supply new tools is very promising to fulfil these requirements.

More advanced devices based on lab-on-chip concepts are very promising tools for controlling cell behavior in vitro in order to mimic in vivo 3D dynamic cell microenvironments and will certainly contribute to the advent of a new generation of cell culture models [131, 132]. Nevertheless, this development must be made while keeping in mind that the system should be reproducible and easily handled in a biology laboratory [133].

Acknowledgements This work was supported by Action 4,221: "Nanobiotechnologies for Health" of the Joint Research Centre. We thanks Dr S. Belkin for the bacteria culture experiments.

References

1. Ratner BD, Bryant S (2004) Biomaterials: where we have been and where. we are going. Annu Rev Biomed Eng 6:41–75
2. Kasemo B (2002) Biological surface science. Surf Sci 500:656–677
3. Langer R, Tirrell DA (2004) Designing materials for biology and medicine. Nature 428:487–492
4. Bhadriraju K, Chen CS (2002) Engineering cellular microenvironments to improve cell-based drug testing. Drug Disc Today 7:612–620
5. Hartung T (2007) Food for thought on cell culture. Altex 24(3):143–147
6. Flaim CJ, Chien S, Bhatia SN (2005) An extracellular matrix microarray for probing cellular differentiation. Nat Methods 2:119–125
7. Anderson DG, Levenberg S, Langer R (2004) Nanoliter-scale synthesis of arrayed biomaterials and application to human embryonic stem cells. Nat Biotechnol 22:863–866
8. Chen CS, Mrksich M, Huang S, Whitesides GM, Ingber DE (1997) Geometric control of cell life and death. Science 276:1425–1428
9. Thery M, Racine V, Pepin A, Piel M, Chen Y, Sibarita JB, Bornens M (2005) The extracellular matrix guides the orientation of the cell division axis. Nat Cell Biol 7:947–953
10. Thery M, Racine V, Piel M, Pepin A, Dimitrov A, Chen Y, Sibarita JB, Bornens M (2006) Anisotropy of cell adhesive microenvironment governs cell internal organization and orientation of polarity. Proc Natl Acad Sci U S A 103(52):19771–19776
11. Khetani SR, Bhatia SN (2007) Microscale culture of human liver cells for drug development. Nat Biotechnol Lett 2007:1–7
12. Ziegler C (2000) Cell-based biosensors. Fresenius J Anal Chem 366:552–559
13. Elad T, Lee JH, Belkin S, Gu MB (2008) Microbial whole-cell arrays. Microb Biotechnol 1(2):137–148
14. Garcia AJ (2006) Interfaces to control cell-biomaterial adhesive interactions. Adv Polym Sci 203:171–190
15. Arnold M, Cavalcanti-Adam A, Glass R, Blummel J, Eck W, Kessler H, Spatz JP (2004) Activation of integrin function by nanopatterned adhesive interfaces. Chem Phys Chem 3:383–388
16. Mrksich M (2002) What can surface chemistry do for cell biology? Curr Opin Chem Biol 6:794–797
17. Underwood PA, Steel JG, Dalton BA (1993) Effects of polystyrene surface chemistry on the biological activity of solid phase fibronectin and vitronectin, analysed with monoclonal antibodies. J Cell 104(3):793–803
18. Andrade JD, Hlady VL, Wagenen RAV (1984) Effects of plasma protein adsorption on protein conformation and activity. Pure Appl Chem 56:1345–1350
19. Wertz CF, Santore MM (1999) Adsorption and relaxation kinetics of albumin and fibrinogen on hydrophobic surfaces: single-species and competitive behavior. Langmuir 15:8884–8894
20. Love JC, Estroff LA, Kriebel JK, Nuzzo RG, Whitesides GM (2005) Self-assembled monolayers of thiolates on metals as a form of nanotechnology. Chem Rev 105:1103–1169
21. Mrksich M, Whitesides GM (1996) Using self assembled monolayers to understand the interactions of made man surfaces with proteins and cells. Annu Rev Biophys Biomol Struct 25:55–78
22. Michael KE, Vernekar VN, Keselowsky BG, Meredith JC, Latour RA, Garcia AJ (2003) Adsorption-induced conformational changes in fibronectin due to interactions with well-defined surface chemistries. Langmuir 19:8033–8040
23. Keselowsky BG, Collard DM, Garcia AJ (2004) Surface chemistry modulates focal adhesion composition and signaling through changes in integrin binding. Biomaterials 25:5947–5954
24. Lan MA, Gersbach CA, Michael KE, Keselowsky BG, Garcia AJ (2005) Myoblast proliferation and differentiation on fibronectin-coated self assembled monolayers presenting different surface chemistries. Biomaterials 26:4523–4531
25. Lan S, Veiseh M, Zhang M (2005) Surface modification of silicon and gold-patterned silicon surfaces for improved biocompatibility and cell patterning selectivity. Biosens Bioelectron 20:1697–1708

26. Wang H, He Y, Ratner BD, Jiang S (2005) Modulating cell adhesion and spreading by control of FnIII7–10 orientation on charged self–assembled monolayers (SAMs) of alkanethiolates. J Biomed Mater Res A 77A(4):672–678
27. Liu L, Chen S, Giachelli C, Ratner B, Jiang S (2005) Controlling osteopontin orientation on surfaces to modulate endothelial cell adhesion. J Biomed Mater Res A 74A:23–31
28. Faucheux N, Schweiss R, Lutzow K, Werner L, Groth L (2004) Self-assembled monolayers with different terminating groups as model substrates for cell adhesion studies. Biomaterials 25:2721–2730
29. Hersel U, Dahmen C, Kessler H (2003) RGD-modified polymers: biomaterials for. stimulated cell adhesion and beyond. Biomaterials 24:4385–4415
30. Roberts C, Chen CS, Mrksich M, Martichonok V, Ingber DE, Whitesides GM (1998) Using mixed self-assembled monolayers presenting RGD and (EG)3OH groups to characterize long-term attachment of bovine capillary endothelial cells to surfaces. J Am Chem Soc 120:6548–6555
31. Houseman BT, Mrksich M (2001) Biomaterials 22:943–955
32. Cavalcanti-Adam EA, Micoulet A, Blummel J, Auernheimer J, Kessler H, Spatz JP (2006) Eur J Cell Biol 85:219–224
33. Castel D, Pitaval A et al. (2006) Cell microarrays in drug discovery. Drug Disc Today 11(13):616–622
34. Ruiz A, Ceriotti L, Buzanska L, Hasiwa M, Bretagnol F, Ceccone G, Gilliland D, Rauscher H, Coecke S, Colpo P, Rossi F (2007) Controlled micropatterning of biomolecules for cell culturing. Microelectron Eng C 84:1733–1736
35. Prime KL, Whitesides GM (1991) Self-assembled organic monolayers: model systems for studying adsorption of proteins at surfaces. Science 252:1164
36. Prime KL, Whitesides GM (1993) Adsorption of proteins onto surfaces containing end-attached oligo(ethylene oxide): a model system using self-assembled monolayers. J Am Chem Soc 115:10714
37. Morra M (2000) J Biomater Sci Polym Ed 11:547
38. Kingshott P, Griesser HJ (1999) Curr Opin Solid State Mater Sci 4:403–412
39. Jeon SI, Lee JH, Andrade JD, de Gennes PG (1991) Colloid Interf Sci 142:159–166
40. Harder P, Grunze M, Dahnit R, Whitesides GM, Laibinis PE (1998) J Phys Chem B 102:426–429
41. Herrwerth S, Eck W, Reinhardt S, Grunze M (2000) J Am Chem Soc 125(31):9359–9366
42. Kenausis GL, Voros J, Elbert DL, Huang NP, Hofer R, Ruiz- Taylor L et al. (2000) J Phys Chem B 104:3298–3309
43. Chapman RG, Ostuni E, Takayama S, Holmlin RE, Yan L, Whitesides GM (2000) Surveying for surfaces that resist the adsoption of proteins. J Am Chem Soc 122:8303–8304
44. Ostuni E, Chapman RG, Holmin RE, Takayama S, Whitesides GM (2001) A survey of structure-property relationships of surfaces that resist the adsorption of protein. Langmuir 17:5605–5620
45. Ostuni E, Chapman RG, Liang MN, Meluleni G, Pier G, Ingber DE, Whitesides GM (2001) Self-assembled monolayers that resist the adsorption of proteins and the adhesion of bacterial and mammalian cells. Langmuir 17:6336–6343
46. Chapman RG, Ostuni E, Liang MN, Meluleni G, Kim E, Yan L, Pier G, Warren HS, Whitesides GM (2001) Polymeric thin films that resist the adsorption of proteins and the adhesion of bacteria. Langmuir 17(4):1225–1233
47. Flynn NT, Tran T, Cima M, Langer R (2003) Long term stability of self- assembled monolayers in biological media. Langmuir 19:10909–10915
48. Jiang X, Bruzewicz DA, Thant MM, Whitesides GM (2004) Palladium as a substrate for self-assembled monolayers used in biotechnology. Anal Chem 76:6116–6121
49. Nelson CM, Raghavan S, Tan JL, Chen CS (2003) Degradation of micropatterned surfaces by cell-dependent and -independent processes. Langmuir 19:1493–1499
50. Ratner BD, Chilkoti A, Lopez GP (1990) In: d'Agostino R (ed) Plasma deposition, treatment and etching of polymers. Academic, San Diego, p 463
51. Sardella E, Favia P, Gristina R, Nardulli M, d'Agostino R (2006) Plasma Process Polym 3:456

52. Bretagnol Fr, Lejeune M, Papadopoulou A, Hasiwa M, Rauscher H, Ceccone G, Colpo P, Rossi F (2006) Fouling and non-fouling surfaces produced by plasma polymerization of ethylene oxide monomer. Acta Biomater 2(2):165–172
53. Bretagnol F, Valsesia A, Ceccone G, Colpo P, Gilliland D, Ceriotti L, Hasiwa M, Rossi F (2006) Plasma Process Polym 3 (6-7): 443–455
54. Bretagnol F, Ceriotti L, Lejeune M, Papadopoulou A, Hasiwa M, Gilliland D, Ceccone G, Colpo P, Rossi F (2006) Functional micropatterned surfaces by combination of plasma polymerization and lift-off processes. Plasma Process Polym 3(1):30–38
55. Siow KS, Britcher L, Kumar S, Griesser HJ (2006) Plasma Process Polym 3:392–418
56. Cho DL, Claesson PM, Goelander CG, Johansson K (1990) Structure and surface properties of plasma polymerized acrylic acid layers. J Appl Polym Sci 41:1373
57. Candan S, Beck AJ, O'Toole L, Short RD (1998) Effects of processing parameters in plasma deposition: acrylic acid revisited. J Vacuum Sci Technol A 16(2):1702–1709
58. Detomaso L, Gristina R, Senesi GS, d'Agostino R, Favia P (2005) Stable plasma-deposited acrylic acid surfaces for cell culture applications. Biomaterials 26:3831
59. Rossini P, Colpo P, Ceccone G, Jandt KD, Rossi F (2002) Surfaces engineering of polymeric films for biomedical applications. Mater Sci Eng C 1050:1
60. Lejeune M, Brétagnol M, Ceccone G, Colpo P, Rossi Fr (2006) Microstructural evolution of allylamine polymerized plasma films. Surf Coat Technol 200(20/21):5902–5907
61. Thissen H, Johnson G, Hartley PG, Kingshott P, Griesser HJ (2006) Two-dimensional patterning of thin coatings for the control o f tissue outgrowth. Biomaterials 27(1):35–43
62. Blättler T, Pasche S, Textor M, Griesser HJ (2006) High salt stability and protein resistance of poly(L-lysine)-*g*-poly(ethylene glycol) copolymers covalently immobilized via aldehyde plasma polymer interlayers on inorganic and polymeric substrates. Langmuir 22:5760–5769
63. Thierry B, Jasieniak M, de Smet LCPM, Vasilev K, Griesser HJ (2008) Reactive epoxy-functionalized thin films by a pulsed plasma polymerization process. Langmuir 24(18):10187–10195
64. Lopez GP, Ratner BD, Tidwell CD, Haycox CL, Rapoza RJ, Horbett TA (1992) Glow discharge plasma deposition of tetraethylene glycol dimethyl ether for fouling-resistant biomaterial surfaces. J Biomed Mater Res 26:415
65. Haddow DB, Steele DA, Short RD, Dawson RA, Macneil S (2003) Plasma-polymerized surfaces for culture of human keratinocytes and transfer of cells to an in vitro wound-bed model. J Biomed Mater Res A 64A(1):80–87
66. Sardella E, Gristina R, Ceccone G, Gilliland DP, Apadopoulou-Bouraoui A, Rossi F, Senesi GS, Detomaso L, Favia P, d'Agostino R (2005) Control of cell adhesion and spreading by spatial microarranged PEO-like and pdAA domains. Surf Coat Technol 200:51
67. Johnston EE, Bryers JD, Ratner BD (2005) Plasma deposition and surface characterization of oligoglyme, dioxane, and crown ether nonfouling films. Langmuir 21:870–881
68. Bretagnol F, Kylian O, Hasiwa M, Ceriotti L, Rauscher H, Ceccone G, Gilliland D, Colpo P, Rossi F (2007) Sens Actuators B 123:283–292
69. Falconnet D, Csucs G, Grandin HM, Textor M (2006) Surface engineering approaches to micropattern surfaces for cell-based assays. Biomaterials 27:3044–3063
70. Vörös J, Blättler T, Textor M (2005) Bioactive patterns at the 100 nm scale produced via multifunctional physisorbed adlayers. MRS Bull 30(3):202–206
71. Slater JH, Frey W (2008) Nanopatterning of fibronectin and the influence of integrin clustering on endothelial cell spreading and proliferation. J Biomed Mater Res A 87A(1):176–195
72. Blawas AS, Reichert WM (1998) Protein patterning. Biomaterials 19:595–609
73. Iwanaga S, Akiyama Y, Kikuchi A, Yamato M, Sakai K, Okano T (2005) Fabrication of a cell array on ultrathin hydrophilic polymer gels utilising electron beam irradiation and UV excimer laser ablation, Biomaterials 26:5395–5404
74. Nakayama H, Kikuchi Y, Takarada T, Nakayama H, Yamaguchi K, Maeda M (2006) Spatiotemporal control of cell adhesion on a self-assembled monolayer having a photocleavable protecting group. Anal Chim Acta 578:100–104

75. detrait E, Lhoest J-B, Knoops B, Bertrand P, van den Bosch, de Aguilar P (1998) Orientation of cell adhesion and growth on patterned heterogeneous polystyrene surface J Neurosci Methods 84:193–204
76. Mikulikova R, Moritz S, Gumpenberger T, Olbrich M, Romanin C, Bacakova L, Svorcik V, Heitz J (2005) Cell microarrays on photochemically modified polytetrafluoroethylene Biomaterials 26:5572–5580
77. Schena M, Shalon D, Davis RW, Brown PO (1995) Quantitative monitoring of gene expression patterns with a complementary DNA microarray. Science 270:467–470
78. Fodor SP, Rava RP, Huang XC, Pease AC, Holmes CP, Adams CL (1993) Multiplexed biochemical assays with biological chips. Nature 364:555–556
79. Revzin A, Russel RJ, Yadavalli VK, Koh WG, Deister C, Hile DD, Mellott MB, Pishko MV (2001) Fabrication of poly(ethylene glydol) hydrogel microstructures using photolithography. Langmuir 17:5440–5447
80. Hahn MS, Taite LJ, Moon JJ, Rowland MC, Ruffino KA, West JL (2006) Photolithographic patterning of polyethylene glycol hydrogels. Biomaterials 27:2519–2524
81. Revzin A, Tompkins RG, Toner M (2003) Surface engineering with poly(ethylene glycol) photolithography to create high-density cell arrays on glass. Langmuir 19:9855–9862
82. Goessl A, Garrison MD, Lhoest JB, Hoffman AS (2001) Plasma lithography – thin-film patterning of polymeric biomaterials by RF plasma polymerization I: Surface preparation and analysis. J Biomater Sci Polym Ed 12:721–738
83. Favia P, Sardella E, Gristina R, d'Agostino R (2003) Novel plasma processes for biomaterials: micro-scale patterning of biomedical polymers. Surf Coat Technol 169:707–711
84. Xia Y, Whitesides GM (1998) Soft lithography. Angew Chem Int Ed Engl 37:550–575
85. Whitesides GM, Ostuni E, Takayama S, Jiang X, Ingber DE (2001) Soft lithography in biology and biochemistry. Annu Rev Biomed Eng 3:335–373
86. Zhao XM, Xia Y, Whitesides GM (1996) Fabrication of three-dimensional microstructures: microtransfer molding. Adv Mater 8:837–840
87. Xia Y, Kim E, Zhao XM, Rogers JA, Prentiss M, Whitesides GM (1996) Complex optical surfaces by replica molding against elastomeric masters. Science 273:347–349
88. Xia Y, McClelland JJ, Gupta R, Qin D, Zhao XM, et al. (1997) Replica molding using polymeric materials: a practical step toward nanomanufacturing. Adv Mater 9:147–149
89. Singhvi R, Kumar A, Lopez GP, Stephanopoulos GN, Wang DIC, Whitesides GM (1994) Engineering cell shape and function. Science 264:696–698
90. Bernard A, Delamarche E, Schmid H, Michel B, Bosshard HR, Biebuyck H (1998) Printing patterns of proteins. Langmuir 14:2225–2229
91. Michel B, Bernard A, Bietsch A, Delamarche E, Geissler M, Juncker D, Kind H, Renault JP, Rothuizen H, Schmid H, Schmidt-Winkel P, Stutz R, Wolf H (2001) Printing meets lithography: soft lithography approaches to high-resolution patterning. IBM J Res Dev 45:697–719
92. Kane RS, Takayama S, Ostuni E, Ingber DE, Whitesides GM (1999) Patterning proteins and cells using soft lithography. Biomaterials 20:2363–2376
93. Delamarche E (2004) Microcontact printing of proteins. Nanobiotechnology, Chap 3. Wiley, New York
94. Raghjavan S, Chen CS (2004) Micropatterned environments in cell biology. Adv Mater 16:1303–1313
95. Shen CJ, Fu J, Chen CS (2008) Pattening cell and tissue function. Cell Mol Bioeng 1:15–23
96. Kumar G, Ho CC, Co CC (2007) Guiding cell migration using one-way micropattern arrays. Adv Mater 19:1084–1090
97. Dike L, Chen C, Mrksich M, Tien J, Whitesides G, Inger D (1999) Geometric control of switching between growth, apoptosis, and differentiation during angiogenesis using micropatterned substrates. In vitro cell. Dev Biol 35:441–448
98. Lehnert D, Wehrle-Haller B, David C, Weiland U, Ballestrem C, Imhof BA, Bastmeyer M (2004) Cell behaviour on micropatterned substrata: limits of extracellular matrix geometry for spreading and adhesion. J Cell Sci 117:41–52

99. Ruiz A, Buzanska L, Gilliland G, Rauscher H, Sirghi L, Sobanski T, Zychowicz Marzena, Ceriotti L, Bretagnol F, Coecke S, Colpo P, Rossi F (2008) Micro-stamped surfaces for the patterned growth of neural stem cells. Biomaterials 29(36):4766–4774
100. Ruiz A, Buzanska L, Gilliland D, Rauscher H, Sirghi L, Sobanski T, Zychowicz M, Ceriotti L, Bretagnol F, Coecke S, Colpo P, Rossi F (2008) Micro-stamped surfaces for the patterned growth of neural stem cells. Biomaterials 29:4766–4774
101. Pan V, McDevitt TC, Kim TK, Leach-Scampavia D, Stayton PS, Denton DD, Ratner BD (2002) Micro-scale cell patterning on nonfouling plasma polymerized tetraglyme coatings by protein microcontact printing. Plasma Polym 7:171–183
102. Delamarche E, Donzel C, Kamounah FS, Wolf H, Geissler M, Stutz R, Schmidt-Winkel P, Michel B, Mathieu HJ, Schaumburg K (2003) Microcontact printing using poly(dimethylsiloxane) stamps hydrophilized by poly(ethylene oxide) silanes. Langmuir 19:8749–8758
103. Delamarche E, Donzel C, Kamounah FS, Wolf H, Geissler M, Stutz R, Schmidt-Winkel P, Michel B, Mathieu HJ, Schaumburg K (2003) Langmuir 19:8749–8758
104. Delamarche E, Donzel C, Kamounah FS, Wolf H, Geissler M, Stutz R et al. (2003) Microcontact printing using poly(dimethylsiloxane) stamps hydrophilized by poly(ethylene oxide) silanes. Langmuir 19:8749–8758
105. Jackman RJ, Duffy DC, Cherniavskaya O, Whitesides GM (1999) Patterning electroluminiscent materials at feature sizes as small as 5 μm using elastomeric membranes as masks for dry liftoff. Adv Mater 11:546–552
106. Kenis PJA, Ismagilov RF, Whitesides GM (1999) Microfabrication inside capillaries using multiphase laminar flow patterning. Science 285:83–85
107. Delamarche E, Bernard A, Schmid H, Michel B, Biebuyck H (1997) Patterned delivery of immunoglobulins to surfaces using microfluidic networks. Science 276:779–781
108. Chiu DT, Jeon NL, Huang S, Kane RS, Wargo CJ et al. (2000) Patterned deposition of cells and proteins onto surfaces by using three-dimensional microfluidic systems. Proc Natl Acad Sci U S A 97:2408–2413
109. Bhatia SN, Balis UJ, Yarmush ML, Toner M (1998) Microfabrication of hepatocyte/fibroblast co-cultures: role of homotypic cell interactions. Biotechnol Prog 14:378–387
110. Fukuda J, Khademhosseini A, Yeh J, Eng G, Cheng J, Farokzad OC, Langer R (2006) Micropatterned cell co-cultures using layer-by-layer deposition of extracellular matrix components. Biomaterials 27:1479
111. Yousaf MN, Houseman BT, Mrksich M (2001) Using electroactive substrates to pattern the attachment of two different cell populations. Proc Natl Acad Sci U S A 98:5992–5996
112. Jiang X, Ferrigno R, Mrksich M, Whitesides GM (2003) Electrochemical desorption of self-assembled monolayers noninvasively releases patterned cells from geometrical confinements. J Am Chem Soc 125:2366–2367
113. Tsuda Y, Kikuchi A, Yamato M, Nakao A, Sakurai Y, Umezu M et al. (2005) The use of patterned dual thermoresponsive surfaces for the collective recovery as co-cultured cell sheets. Biomaterials 26:1885–1893
114. Yamato M, Konno C, Utsumi M, Kikuchi A, Okano T (2002) Thermally responsive polymer-grafted surfaces facilitate patterned cell seeding and co-culture. Biomaterials 23:561–567
115. Suh KY, Khademhosseini A, Yoo PJ, Langer R (2004) Patterning and separating infected bacteria using host-parasite and virus-antibody interactions. Biomed Microdev 6:223–229
116. Suh KY, Khademhosseini A, Yang JM, Eng G, Langer R (2004) Soft lithographic patterning of hyaluronic acid on hydrophilic substrates using molding and printing. Adv Mater 16:584–588
117. Bjerketorp J, Kansson SH, Belkin S, Jansson JK (2006) Advances in preservation methods: keeping biosensor microorganisms alive and active. Curr Opin Biotechnol 17:43–49
118. Razatos A, Ong YL, Sharma MM, Georgiou G (1998) Molecular determinants of bacterial adhesion monitored by atomic force microscopy. Proc Natl Acad Sci U S A 95:11059–11064
119. Xu L, Robert L, Ouyang Q, Taddei F, Chen Y, Lindner AB, Baigl D (2007) Microcontact printing of living bacteria arrays with cellular resolution. Nanoletters 7:2068–2072

120. Saravia V, Kupcu S, Nolte M, Huber C, Pum D, Fery A, Sleytr UB, Toca-Herrera JL (2007) Bacterial patterning by micro-contact printing of PLL-g-PEG. J Biotechnol 130:247–252
121. Weibel DB, Lee A, Mayer M, Brady SF, Bruzewicz D, Yang J, DiLuzio WR, Clardy J, Whitesides GM (2005) Bacterial printing press that regenerates its ink: contact-printing bacteria using hydrogel stamps. Langmuir 21:6436–6442
122. Shim HW, Lee JH, Hwang TS, Rhee YW, Bae YM, Choi JS, Han J, Lee CS (2007) Patterning of proteins and cells on functionalized surfaces prepared by polyelectrolyte multilayers and micromolding in capillaries. Biosens Bioelectron 22:3188–3195
123. Howell SW, Inerowicv HD, Regnier FE, Reifenberg R (2003) Patterned protein microarrays for bacterial detection. Langmuir 19:436–439
124. Suo Z, Avci R, Yang X, Pascal DW (2008) Efficient immobilization and patterning of live bacterial cells. Langmuir 24:4161–4167
125. Yi DK, Kim MJ, Turner L, Breuer KS, Kim DY (2006) Colloid lithography-induced polydimethylsiloxane microstructures and their application to cell patterning. Biotechnol Lett 28:169–173
126. Akselrod GM, Timp W, Mirsaidov U, Zhao Q, Li C, Timp R, Matsudaira P, Timp G (2006) Laser-guided assembly of heterotypic three-dimensiona living cell microarrays. Biophys J 91:3465–3473
127. Kuschel C, Steuer H, Maurer AN, Kanzok B, Stoop R, Angres B (2006) Cell adhesion profiling using extracellular matrix protein microarrays. BioTechniques 40:523–530
128. Soen Y, Mori A, Palmer TD, Brown PO (2006) Exploring the regulation of human neural precursor cell differentiation using arrays of signalling microenvironments. Mol Syst Biol 2:37
129. Griffith LG, Swartz MA (2006) Capturing complex 3D tissue physiology in vitro. Nat Rev Mol Cell Biol 7:211–224
130. Cukierman E, Pankov R, Stevens DR, Yamada KM (2001) Taking cell-matrix adhesions to the third dimension. Science 23(294):1708–1712
131. El-Ali1 J, Sorger PK, Jensen KF (2006) Cells on chips. Nature 442(27):403–411
132. Yeo WS, Yousaf MN, Mrksich MJ (2003) Dynamic interfaces between cells and surfaces: electroactive substrates that sequentially release and attach cells. Am Chem Soc 125(49):14994–14995
133. Fink J, Thery M et al. (2007) Comparative study and improvement of current cell micro-patterning techniques. Lab Chip 7:672–680
134. Anderson JR, Chiu DT, Jackman RJ, Cherniavskaya O, McDonald JC et al. (2000) Fabrication of topologically complex three-dimensional microfluidic systems in PDMS by rapid prototyping. Anal Chem 72:3158–3164
135. Latour RA (2005) Encyclopedia of Biomaterials and Biomedical Engineering, ed. G.L.B.G. Wnek. New York: Taylor & Francis
136. Pierschbacher MD, Ruoslahti E (1984) The cell attachment activity of fibronectin can be duplicated by small fragments of the molecule. Nature 309:30–33
137. Geissler and Xia (2004) Patterning: Principles and Some New Developments, Advanced materials, vol.16, Issue 15, Pages 1249–1269
138. Albrecht DR, Underhill GH, Wassermann TB, Sah RL, Bhatia SN. (2006) Probing the role of multicellular organization in 3D microenvironments. Nature Methods 3, 369–375
139. Cerotti L, Buzanska L, Rauscher H, Mannelli I, Sirghi L, Gilliland D, Hasiwa M, Bretagnol F, Zychowicz M, Ruiz A, Bremer S, Coecke S, Colpo P and Rossi F (2009) Fabrication and characterization of protein arrays for stem cell patterning. Soft Matter 5:1406–1416

Adv Biochem Engin/Biotechnol (2010) 117: 131–154
DOI:10.1007/10_2009_6

Published online: 5 June 2009

Fiber-Optic Based Cell Sensors

Evgeni Eltzov and Robert S. Marks

Abstract Different whole cell fiber optic based biosensors have been developed to detect the total effect of a wide range of environmental pollutants, providing results within a very short period. These biosensors are usually built from three major components, the biorecognition element (whole-cells) intimately attached to a transducer (optic fiber) using a variety of techniques (adsorption, covalent binding, polymer trapping, etc). Even with a great progress in the field of biosensors, there is still a serious lack of commercial applications, capable of competing with traditional analytical tools.

Keywords bioreporter • bioluminescence • fiber optic biosensors

Contents

E. Eltzov
Unit of Environmental Engineering, Faculty of Engineering Science,
Ben-Gurion University of the Negev, Beer-Sheva, Israel

R.S. Marks (✉)
Department of Biotechnology Engineering, Faculty of Engineering Science,
Ben-Gurion University of the Negev, Beer-Sheva, Israel
National Institute for Biotechnology in the Negev Ben-Gurion University of the Negev,
Beer-Sheva, Israel
e-mail: rsmarks@bgu.ac.il

1 Introduction

Seriousness of environmental problems fuels a growing number of initiatives and law-making actions to control pollution. Each year numerous new compounds, with unknown effects on human health, have been developed and eventually found their way into the environment. For example, the environmental protection agency (EPA) reported that 141 unregulated chemicals were found in tap water in 45 states in the US, 40 of which were served to at least one million people [1]. Thus, there is a huge demand for fast and cost-effective analytical techniques to monitor wider ranges of analytes in air, water and soil, and to do so with greater frequency and accuracy. Conventional techniques are based on chemical or physical analyses and allow highly accurate and sensitive determination of the exact composition of any sample [2]. However, these methods have four main disadvantages. The first is that these techniques, e.g., liquid chromatography (HPLC), gas chromatography (GL) or enzyme-linked immunosorbent assay (ELISA) [3], enable just the detection of a single compound or a group of structurally related compounds at any given time. The second disadvantage is that these methods provide no indication about the biological effects of the target compound [4], while many various compounds with dissimilar chemical structures may have the same biological toxic effect [5]. Third, all these techniques require well skilled personnel and expensive equipment, and finally, results of these tests may take a few days to a few weeks. To help overcome these problems, the use of alternative biomethods was suggested.

Water pollution is of great concern in modern society and is often monitored using aquatic organisms with on line applications being favored [6]. For this purpose, different aquatic microorganisms were used: e.g., algae [7-10], *Daphnia* [9, 10] and various types of fish [11-13]. These biomonitors detect the total effect of herbicides and heavy metals in real-time, providing fast alarms when a contamination peak occurs [14-16]. However, the effect on these organisms has no clear link with the hazard for humans, and these biomonitors do not detect all important toxic contaminants to humans (e.g., genotoxicants and endocrine disruptive compounds (EDCs)), because they only detect acute toxicity and not chronic toxicity. The last decade witnessed an extraordinary growth in research on sensors in general and whole cell biosensors in particular. Many researchers have been involved in developing and applying bioassay systems, which use genetically engineered whole cells, for the toxicity testing of water, sediment, and soil samples. This type of testing has been in use for many years but suffers from deficiencies of cost per test, time to obtain test results (samples need to be brought to the lab), and inherent variability of the test data [17]. To solve these problems, these bioassays have been adapted to biosensor transducers that use the engineered biological material as an analytical tool. Also, unlike biosensors, bioassays or bioanalytical systems require additional processing steps, such as a reagent addition [18]. Various whole cell bacterial strains have been developed to detect the total effect of contaminants causing human toxicity, e.g., genotoxicity, membrane damage, oxidative damage and protein damage, providing results in 1-2 h. These bacteria are genetically modified organisms (GMOs), engineered to luminesce after exposure to certain toxic compounds. Luminescence is easily measured using a photodetector and does not suffer from a variable background signal a from water matrix. For example, responses in bacteria of oxidative (peroxide based) stress confirm the oxidative nature of the toxicant. Moreover, in many cases, convincing correlations have been reported between the results obtained using microbial tests and those derived from long term assays using higher organisms [19–21]. Biosensors allow discovery, detection and prediction of biological effects (toxicity) of various contaminants in water, air and soil samples.

2 Whole Cell Fiber Optic Biosensor

2.1 Biosensors as an Alternative Bioassay

A generally accepted definition of a biosensor, is that of a self contained, bionic, integrated device that includes a biological recognition element (e.g., microorganisms) that can respond in a concentration dependent manner to a biochemical species measurand [22]. The characteristic biosensor structure integrates: a bio recognition component, immobilized to an interface surface of a transducing element [4]. The biological recognition elements of a biosensor interact selectively with the target analyte(s), assuring the selectivity of the sensors. These elements can be classified into five main classes: whole cells, nucleic acids, immunochemicals, enzymatic and non enzymatic receptors [23]. The immobilization strategy depends on the bioreceptor and type of transducer. Some conditions that must be considered are (1) maintaining biological activity after immobilization, (2) proximity of the biological layer to the transducer, (3) stability and durability of the biological layer, (4) sensing specificity of the biological component to its analyte [24], and (5) for some uses, the possible future reuse of some biomaterials [25]. The principle immobilization methods are adsorption, cross linking, covalent binding, entrapment, sol–gel entrapment and Langmuir–Blodgett (LMB) deposition of self-assembled biomembranes [24].

Biosensors can be classified not only by their biorecognition elements but also by the transducing methods they employ. There are four major groups: electrochemical, optical, mass sensitive, and thermal. Optical transducers offer the largest number of possible detection strategies and may use the following techniques, e.g., UV–Vis absorption, bioluminescence, chemiluminescence, fluorescence, phosphorescence, surface plasmon resonance, evanescent wave spectroscopy, reflectance, scattering and refractive index changes produced by the interaction of the receptor with the target analyte [24, 26]. Some of these techniques have certain advantages in that they are simple, flexible and allow for multichannel and remote sensing.

2.2 Fiber Optics as Ideal Transducers

Optical Fibers may act as transduction elements, in detecting target biomolecules [27]. They are ideal transducers governed by Snell's law, having the following advantages: (1) geometric convenience and flexibility (2) low cost of production,

(3) are inert, and therefore nonhazardous and biocompatible, (4) being non electrical are thus free of signal interference, (5) being dielectric they are protected from atmospheric disturbance, (6) their small volume economizes reagents and enables convenient portability or storage as well as access to difficult areas, (7) are robust with high tensile strength, (8) their silica composition enables chemical modification when required, (9) they enable solid phase characterization of the analyte, (10) their potentially long interaction lengths enable remote signal transmission, (11) light transmission occurs with minimal loss, (12) enable high efficiency coupling in the blue–green region which is ideal for bioluminescence, (13) exhibit optical multiplicity allowing them to be used in other systems such as chemiluminescence and fluorescence, (14) optrode systems are polyvalent as they may be easily adapted from one whole cell reporter system analyte to another, (15) are amenable to mass production, and (16) can transmit multiple optical signals simultaneously, thereby offering multiplexing capabilities for sensing [28, 29].

Usually, fiber optics are made out of glass or plastic and have several possible configurations, formats, shapes, and sizes. Rapid progress in telecommunication applications expanded usage of these new, less expensive and more advanced fiber optics in the biosensors field. Optical fibers are built from three parts: a core with a refractive index, n_1, a cladding with a lower refractive index, n_2, and a jacket for protection from environmental stress (Fig. 1). There are two basic conditions for light propagation in a fiber. First, light should strike the cladding at an angle greater than the critical angle (φ_c) and second, angles of the light entering the fiber should be within the acceptance cone. When the light angle is less than the critical angle (φ_c) it will be both partially reflected and refracted. Glass fibers are the most commonly used in optrode

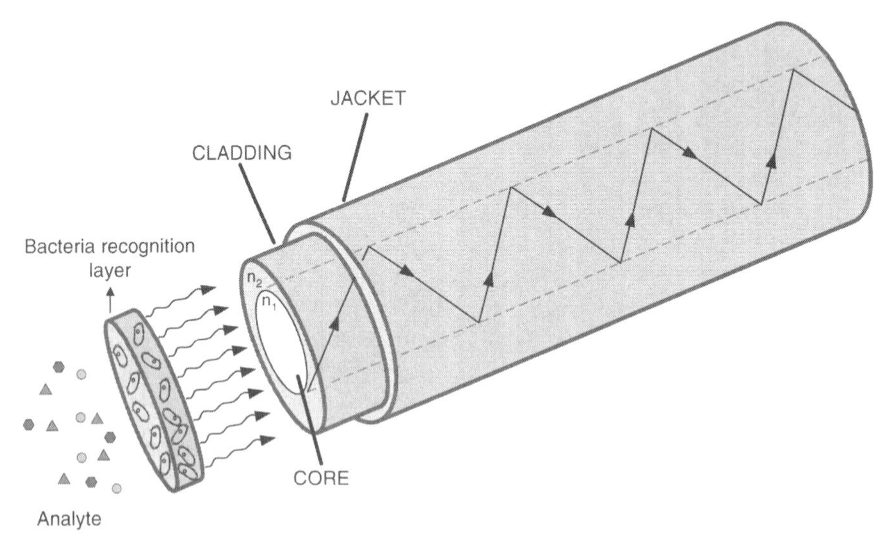

Fig. 1 Schematic representation of a bacterial fiber optic. Only when interface between the core ($n1$) is bigger that clad, ($n2$) ($n1 > n2$) will the light propagation through the optical fiber

Table 1 Advantages and disadvantages of fiber optic biosensors

Advantages		Disadvantages
Not subjected to electrical interferences		Light interference
Repairability (immobilized biocomponent does not have to be in contact with fiber optic or transducing element)		Slow response time (dependency on the analyte mass transfer to the bacteria)
Low-loss (allow remote monitoring environmental samples)	⟷	Limited stability of the immobilized biological component
Simple		
No references electrode		Irreversibility (problem if the reagent is expensive)
Low-cost		

biosensors. The possibility to transfer light in the visible and near infrared regions of the optical spectrum (400 nm $< \lambda <$ 700 nm) make some types of fiber suitable for measuring luminescent signals generated by whole cell organisms [30].

The optical imaging fiber optic bundles make use of the ability of the fiber to carry images from one end to the other, due to their coherent nature. This imaging capability can be utilized to image simultaneously and measure local analyte concentrations with micron-scale resolution [30]. The distal face of an imaging fiber is coated with an analyte sensitive layer, (usually a biorecognition molecule or living cells), which produces a microsensor array spatially capable of resolving analyte concentrations [30].

Table 1 shows the advantages and disadvantages of fiber optic bioluminescence methods employed in sensing various environmental pollutants over classical bioassays in general.

2.3 Immobilization Matrices

Supporting matrices have to prevent bacteria from being washed out from their immobilization matrix and still enable signal transduction and rapid communication with the environment [31]. The immobilization matrix should exhibit many characteristics, e.g., gentle hydration and high porosity, an ability to survive harsh water conditions (i.e., pH, turbidity), allow an inflow of nutrients, oxygen and analytes as well as an outflow of wastes and signals. There are several types of matrices used for whole cells immobilization, e.g., hydrogels, sol–gels or photopolymers. Hydrogels are networks of water insoluble polymer chains produced as a response to various triggers, e.g., ions, heat, light, or other chemicals that would also act as electrophiles. The easily controlled diffusion, high water content, pliability and biocompatibility make these ideal matrices [29, 32–34]. During the immobilization step some cross linking processes expose the bacteria to damaging agents stressing bacteria (such as excessive heating or ultraviolet light), affecting them before being exposed to the tested samples, which may induce false signals or even irreversibly damage them. Therefore, this step should be carefully planned by choosing the right polymer.

A commonly used class of hydrogel immobilization matrices is alginates. Alginates are unbranched polysaccharides produced by marine brown algae and by some bacteria. They consist of 1,4-linked β-D-mannuronic (M) and α-L-guluronic (G) acid residues in different sequences and proportions [35]. The physical properties of alginates depend on the sequence of M and G residues, as well as on the average molecular weights and the molecular weight distribution of the polymer [36]. In the presence of divalent ions such as calcium, alginates spontaneously form gels in a single step process. The technical success of using alginates for entrapment and encapsulation may be attributed to the gentle environment provided by the gels as well as the high porosity provided by the open lattice structure [34]. While some studies used untreated alginate [29], various chemical modifications of alginates for different purposes have been proposed. In particular, carboxylic groups have been used for the following purposes: (1) to couple the alginate to a short peptide (GRGDY) creating an adhesive hydrogel substrate for cultivating anchorage dependent mammalian cells (myoblasts) and for the expression of a differentiated phenotype [37], (2) to couple galactose moieties to the alginate such as ASGP-R ligands to improve anchorage and the interaction of hepatocytes with the alginate, enhancing the functions of the encapsulated hepatocytes in a three-dimensional culture [38], (3) to cross link covalently alginate chains with poly(ethylene glycol) diamines, and to study the mechanical properties of newly modified alginates [39], and (4) to provide the alginate with a new conjugation property, alginate–biotin spheres created by linking biotin to the carboxylic residues of alginate (Fig. 2). However, all publications published to date in the field have shown that there is never destruction of the integrity of these matrices [40]. Coupling of biotin to the alginate was achieved by using aqueous-phase carbodiimide activation chemistry, followed by the binding of biotin hydrazide [34]. The instability in calcium-poor solutions and deterioration in the presence of phosphate and other calcium chelators are putative problems with alginate matrices. The low deformation resistance and the biodegradability of most soft gels are additional incentives to search for alternative encapsulating procedures.

Photopolymers are another class of immobilization matrices. Most photopolymers use visible or ultraviolet light to cross link the monomers used in the formation of the matrices. Some photopolymers utilize harsh chemical initiators to facilitate polymerization. A photon from the light source breaks the photoinitiator into groups of highly energized radicals. The radicals then react with the resident monomer in solution and initiate the usually unstable thermoset polymerization [41] This method may be used as a structural reinforcement of alginate. It was demonstrated that photopolarization increases the strength of alginate immobilization matrices by more than twice its innate strength [42].

Another class of immobilization matrices is sol–gel silicates. Sol–gels are hybrid organic inorganic compounds that bridge between glasses and polymers [43]. Structure and thermal stability, rigidity and transparency make sol–gels' techniques

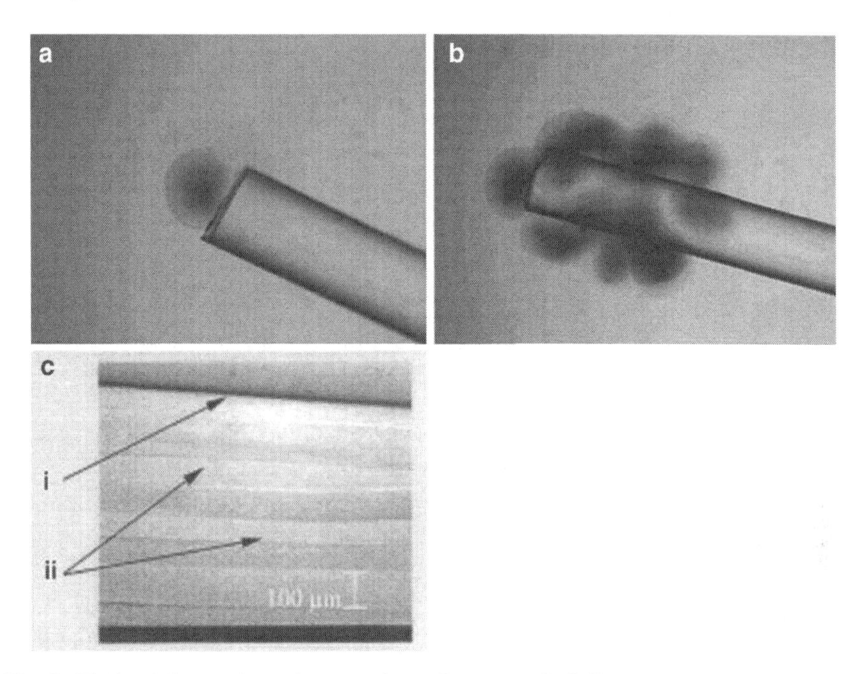

Fig. 2 Biotin–alginate microspheres conjugated to an optical fiber via avidin–biotin affinity interactions. **a)** Attachment of a lone bead to the end of the fiber. **b)** Coating of the fiber with a number of microspheres. Diameter of the optical fiber is in both cases 1000 μm [34]. **c)** Micrograph of probe adlayers set onto the optical fiber core. (i) The fiber–probe interface. (ii) The polymer layers with the approximate thickness at around 80–100 μm [29]

suitable for luminescent bacteria immobilization. Recently, immobilization of viable bacterial cells was successfully performed using silica gel formation at room temperature [44]. It has been shown that the presence of glycerol during immobilization increases the overall cell viability [45]. Alginate–silica gel combinations were proposed for use as reinforced coatings layers over alginate beads [46].

There are many other approaches for the immobilization of bacteria on solid substrates, such as is done with antibodies [47]. The technology for bonding antibodies on different flat or porous substrates is very versatile and well developed, and it is often used as a first step in the construction of complex sensing elements. The successful combination of this generic approach with reporting microorganisms paves the way for their incorporation on or in virtually any substrate [47]. Several successful reports have been published about encapsulated bacteria in agarose [48, 49]. An immobilization process based on the cross linking of bifunctional reagents like glutaraldehyde has been successfully carried out in various supports. Although this technique obviates some of the limitations due to covalent binding, the chemical cross linking reagents used are often detrimental to cell viability [48].

So, despite the fact that this section was to address mostly the immobilization of bioluminescence bacteria on different surfaces, it also demonstrated how these integrate recombinant bacteria on a fiber optic.

2.4 Whole Cell Organisms

In light of the obvious need for "real-time" toxicity monitoring, the last few years have seen significant advances in the use of microbes as test organisms. The advantages offered by microbial toxicity testing include high sensitivity, low cost, large test populations and, most important, rapid responses [50]. Some approaches use the organisms modified to overexpress specific enzymes involved in the analytical measurement [51]. It is well known that numerous biochemical enzymatic reactions simultaneously catalyze in bacteria. Some of these enzymes were used separately as sensor recognition elements in fiber optic applications [52]. However, when the detection is based on a sequence of multiple enzymatic reactions, bacterial sensors are particularly advantageous. While enzymatic cascade reactions are very difficult and complicated to accomplish ex vivo, an in vivo bacterial enzymatic system easily transforms the analyte into an optically detectable product [30].

The use of bacteria as test organisms has an additional unique advantage that has only recently been recognized, i.e., being amenable to sophisticated genetic engineering. Indeed, bacteria can be "programmed" to respond in a specific manner to particular classes of compounds. An observed response, therefore, may indicate both the existence of a toxicant, as well as, it's nature. The sensing element in these bacteria is often composed of regulatory proteins and promoter sequences of either chromosomal or plasmid DNA [53]. The different regulatory proteins used for recombinant environmental sensing, outlined herein, are similar to those used in other gene expression assays. All these reporters are either detected readily or possess easily measured activity. Just as the specificity of the final construct depends upon the proper selection of the sensing promoter, the facility, sensitivity and degree of resolution of the detection will depend, to a large extent, upon the proper choice of a reporter [54]. Various approaches for the detection of environmental pollutants based on the different bacterial reporter genes were proposed in the last 10 years or so. The reporter methods (e.g., *lacZ* [55], alkaline phosphatase [56]) are simple to perform and do not require very expensive instrumentation. Some of these methods do not possess intrinsic optical properties and therefore cannot be measured with optical fibers. Recently, bioluminescence methods have come to the fore. Here activation of the reporter luciferase genes will emit, a readily detectable light signal which allows the monitoring of bacterial response in real-time, by simple luminometry (e.g., fiber optic) [57]. Luciferase genes are reporter genes widely used in both prokaryotic as well as eukaryotic systems. Firefly luciferase (*Photinus pyralis*) and bacterial luciferases of *Vibrio harveyi* and *Vibrio fischeri* are the most used as reporter genes. These luciferases from different groups have no apparent evolutionary relationships; even the reactions they catalyze are different [57]. The main advantage of the firefly over the bacterial luciferase is quantum yield (i.e., efficient conversion of chemical energy to light during enzymatic catalysis) – 90% compared to 5–10% [58, 59]. However, expression of the whole bacterial luciferase operon produces light without any additions, thereby allowing real-time monitoring of gene expression, whereas the expression of firefly luciferase genes requires externally added substrate (lucefirine) for luminescence. This will complicate the monitoring processes and increase drastically the final price of the developed biosensor. Bacterial luciferase catalyzes the obligatory aerobic oxidation of a reduced flavin mononucleotide and a long chain aldehyde to yield a flavin mononucleotide and the corresponding carboxylic acid, with a side reaction light emission of around 490 nm. Luciferase is encoded by *luxA* and *luxB* of the lux operon, while the synthesis of enzymes for the aldehyde is encoded by *luxCDE* (Fig. 3) [60, 61].

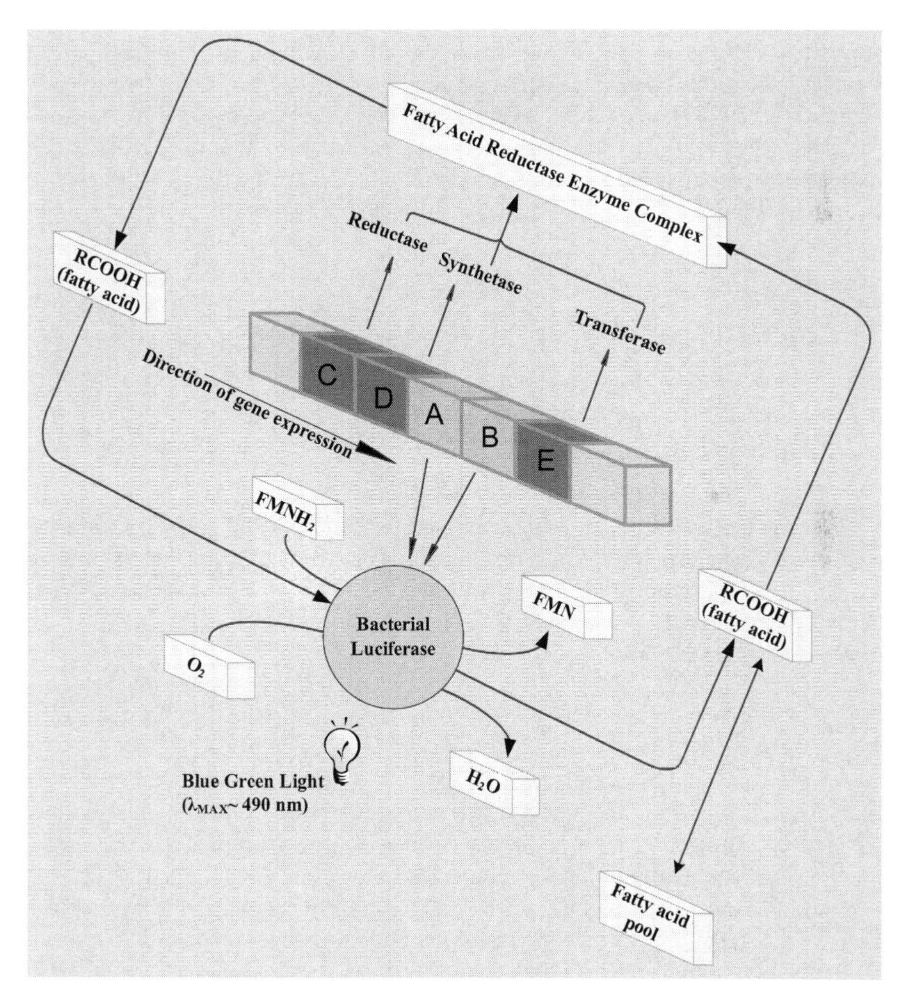

Fig. 3 Relationship of the *lux* genes and corresponding proteins with the bacterial bioluminescence reaction. The fatty acid reductase complex is made up of the three subunits of the reductase, four subunits of the synthetase and four subunits of the transferase. Luciferase is heterodimeric containing one molecule of the α and β subunits each

The detection of toxins using bioluminescent bacterial strains has been broadened as the panel of stress responsive promoters expands [62]. This panel of microbial toxicity sensing systems employs different *Escherichia coli* strains, each carrying a unique stress promoter that activates a different regulatory circuit. It includes the bacterial heat-shock and cold-shock responses to monitor protein damage [47], the SOS regulatory network involved in protection against DNA damage [63, 64], the oxyR and soxRS regulators of oxidative stress response [65], and promoters responsible for bacterial quorum sensing mechanisms [66, 67]. The most significant advantage of this kind of technology is that the bioluminescence response not only indicates the presence of the stress inducing agent but also gives information on its character, namely, its mechanism of action [68]. Moreover, the simple method of using a microtiter plate makes possible the screening of a large number of samples over a short period of time. Current applications of this "panel of stress responsive proteins" include characterization of stress inducing agents [68–71] and the identification and characterization of toxic chemicals [72]. An overview of the analytes or target responses, promoters, and regulatory proteins that mediate reporter gene expression in cell based biosensing systems has been presented [73].

3 Applications

3.1 Fiber Optic Applications Based on Immobilized Microorganisms

Whole cell fiber optic biosensors have been proposed for use in many environmental applications. The capabilities of remote monitoring (fiber tip located at the measurement site and the photodetector far away at distant protected location), small size, durability and flexibility enables the possibility to locate fiber optics in places out of reach for many other biosensors [27].

3.1.1 EDCs

Endocrine disruptive compounds have become a serious problem due to their potential to mimic or antagonize the actions of endogenous hormones at the molecular level. Accumulation of these compounds in mammalian and plant tissues and exposure to humans through the food chains turn them into a real health risk [4]. Polycyclic aromatic hydrocarbons (PAHs) are ubiquitous contaminants, and occur primarily as a result of incomplete combustion processes [74]. The carcinogenic effects of some PAHs are well known, and some have been identified as potential environmental endocrine disruptors. PAHs have two modes of action; by blocking the activation of estrogen receptors, and by induction of Ah-responsive genes that result in a broad spectrum of antiestrogenic responses [75, 76]. Naphthalene and phenanthrene are the most water soluble PAHs, so they are priority pollutants in aqueous solutions. Two different techniques, immobilization of the recombinant bioluminescent bacterial cells and use of a nontoxic biosurfactant, were combined to develop an in situ toxicity biosensor system for phenanthrene detection in soil [77]. Constitutively bioluminescent *E. coli* bacteria were immobilized into a solid agar matrix; furthermore, it was found that the addition of glass beads to the agar media enhanced the stability of the immobilized cells. The biosurfactant, rhamnolipids, was used to extract a model PAH, phenanthrene, and was found to enhance the bioavailability of phenanthrene via an increase in its rate of mass transfer from sorbed soil to the aqueous phase. The monitoring of phenanthrene toxicity was measured by the decrease in bioluminescence when a sample extracted with the biosurfactant was injected into the minibioreactor. The concentrations of phenanthrene in the aqueous phase were correlated well with the corresponding toxicity data obtained by using this biosensor.

Another whole cell biosensor for detection of naphthalene and salicylate was suggested [78, 79]. Using fusion of the nahG gene to the *lux*CDABE operon the reporter *Pseudomonas fluorescens* HK44 cells detected the presence of naphthalene and salicylate in the tested soil. *P. fluorescens* HK44 represents the first genetically engineered microorganism for bioremediation purposes approved for field testing in the United States. The fiber optic tip with immobilized bacteria (using strontium alginate) was placed in a measurement flow cell that received simultaneously a waste stream solution and the maintenance medium. The presence of either naphthalene or salicylate in the waste stream induced rapid bioluminescence signals. The field studies also demonstrated the capability of this sensor to detect the presence of the naphthalene in the contaminated soil. This method promises to be a viable option for in situ determination of environmental contaminant, bioavailability and biodegradation process monitoring and control. Another example of these fiber optic soil biosensors used the genetically engineered *P. putida* strain RB1353, which carries a plasmid borne fusion of the genes for salicylate degradation (*nah*) and luminescence (*lux*) [80]. The relationship between biodegradation and transport of salicylate in porous media and microbial activity was examined. This application consisted of a sample cell, optical fiber, detector, conditioning circuit, and signal processor. The minimum detection level of the naphthalene, in a homogeneous porous medium, was established at 6.7 mg L^{-1}.

EDCs are a class of emerging contaminants that are not defined by their chemical nature but by their biological effect. Therefore the EDC determinations require development of methods based on the monitoring of the biological effect, rather than on chemical analysis. EDCs interfere with endogenous hormone systems, and many of them can bind to the natural estrogen receptor (ER) as agonists or antagonists. This binding ability of ER can be applicable for creating biosensors with natural receptors for testing chemicals with potential environmental toxicity. A biosensor with genetically modified *Saccharomyces cerevisiae* yeast cells entrapped in alginate hydrogel matrices, has been developed and tested for EDC determination [81]. The advantages of this application are as follows. (1) Time - the bioassay was characterized by a total duration time of 2.5 h. In addition it allows for long term storage of the yeast cells, meaning measurements could be made without continuous or repeated cultivation of the cells. (2) Biocompatibility – hydrogels form a protective environment for the entrapped yeast cells, protecting them from contamination, allowing the user to work under non sterile conditions. (3) Simplicity – the bioassay is simple to perform and the hydrogels are both stable and easy to handle. (4) Cost - the method is relatively inexpensive in comparison to LC–MS–MS chemical analysis which requires an expensive instrument and well-trained personnel. (5) Possibility of infield determination of various EDCs in water samples.

A portable optical biosensor, based on a replaceable card with immobilized *E. coli*::*lux* AB TBT3, is described for the on line detection of pollutants [82]. This biosensor was employed for the detection of the well known biocide, tributyltin, mainly used in many fields such as wood treatment and preservation, antifouling of boats (in marine paints), antifungal action in textiles and industrial water systems (cooling tower and refrigeration water systems), wood pulp and paper mill systems, and breweries [4]. The reporter bacteria were immobilized into agarose placed in a parallel epipedic card. For the limitation of photons' dispersion, the space between the bacteria and the fiber optic did not exceed 1 mm. This fiber optic method was able to detect tributyltin as low as 1 nM (325 ng L^{-1}) in an extended time contact of 5 h (Table 2).

Other classes of EDCs are pesticides. The Environmental Protection Agency (EPA) defines a pesticide as "... any substance or mixture of substances intended for preventing, destroying, repelling, or lessening the damage of any pest." Animal studies have shown that prenatal exposure to some pesticides such as methoxychlor or phthalates can reduce spermatogenesis [83, 84]. The pesticide, methyl parathion, was shown to be detectable using a *Flavobacterium sp.* biosensor [85].

Table 2 Examples of different whole cell fibre optic applications for detection of various pollutants

Target	Microorganism	Limit of detection	References
Phenanthrene	*E. coli*	2.06 ppm	[77]
Naphthalene	*Pseudomonas fluorescens* HK4	0.55 mg × L^{-1}	[78]
Salicylate	*Pseudomonas putida*	6.7 mg × L^{-1}	[80]
EDCs (β-E2)	*Saccharomyces cerevisiae*	0.08 mg × L^{-1}	[81]
Tributyltin	*E. coli*	325 ng × L^{-1}	[82]
Methyl parathion	*Flavobacterium sp.*	0.3 μM	[85]
Atrazine	*Scenedesmus subspicatus*	50 ppb	[87]
Herbicides	*Chlorella vulgaris*	0.1 μg L^{-1}	[89]
Hg	*E. coli*	100 nM	[91]
Hg and As	*E. coli*	2.6 μg L^{-1} of Hg(II) and 141 μg L^{-1} of As(V) or 18 μg L^{-1} of As(III)	[92]
Cu	*Alcaligenes eutrophus*	1 μM	[95]
Benzene	*E. coli*	48 ppm	[97]
Mitomycin C	*E. coli*	25 μg L^{-1}	[29]
Glucose	*Staphylococcus warneri*	6 μg L^{-1}	[100]
m-Xylene	*Pseudomonas putida*	5 μmol × L^{-1}	[103]

Toxic compound detection by the use of an immobilized, disposable, microbial glass fiber disc (with immobilized *Flavobacterium sp.* cells) with an optical fiber was a simple, single step and direct measurement of a very low quantity of sample, 75 μL. A lower detection limit of 0.3 μM was estimated from the linear range (4–80 μM) of a calibration plot of an organophosphorus hydrolase enzymatic assay, which was not only better than the reported optical biosensor, but also comparable to the reported amperometric biosensor, for detection of other organophosphate pesticides.

Inhibiting photosynthesis in plants are the most heavily used pesticides in the world – the herbicides 6-chloro-*s*-triazine which include atrazine, cyanazine, propazine and simazine. Under the Food Quality Protection Act (FQPA), triazineherbicides are being considered from the standpoint of endocrine disruption, as well as aggregate and cumulative exposure [86]. An optical biosensor for the determination of herbicides impurities in water, based on immobilized living algae cells, was developed [87]. Most herbicides inhibit the electron transport involved in the photosynthetic processes responsible for ATP production in plants. As a consequence, an increase in chlorophyll fluorescence can be measured and correlated with the concentration of the pollutant. The measuring principle in this research was the determination of the chlorophyll fluorescence depending on the load of water probes, with toxic compounds using fiber optic based electronic equipment. The used microorganisms (*Scenesdesmus subspicatus*) were immobilized on filter paper and covered with alginate. This sensor was able to detect herbicides (e.g., atrazine and endrine) in the $\mu g\ L^{-1}$ concentration range, with 10 min response times. The immobilized organisms can be stored at 4 °C over a period of about 6 months without significant loss in fluorescence properties. This makes the system advantageous for practical applications.

In another method, *Chlorella vulgaris* was used for monitoring herbicides present in aquatic samples [88]. Five membranes, with entrapped algae inside, were placed on a rotating disk, and a fiber optic bundle was located above one of them. While one membrane was exposed to the fiber optic bundle (for 1 min), another four were in the dark (for 4 min). Because membrane bound algal units were refreshed regularly, this application allowed continuous detection of ppb levels of the herbicides. Another *C. vulgaris* application based on kinetic measurements of chlorophyll-a fluorescence in the cells was described [89]. The active membrane was constructed by physical entrapment of algal cells onto a porous matrix with appropriate pore size. Opposite the membrane, a bifurcated bundle of randomized optical fibers was oriented so that the incident light hit the upper part of the membrane and the resulting fluorescence could be collected by the corresponding fibers, and transmitted to the fluorometer equipped with a microcomputer for data recording. The sensor sensitivity was established at 0.1 $\mu g\ L^{-1}$ for a single herbicide, as is required by European Community legislation for drinking water. This device was adequate for the assay under flow conditions, but it could also be used with a single drop of sample deposited on the membrane surface.

3.1.2 Metals

The total amount of metals detected by classical, analytical methods is not always comparable with the actual toxicity of the samples. Therefore, determination of the biologically available metal fraction (bioavailability) in environmental samples is a crucial issue in environmental monitoring [90]. For this issue various fiber optic based methods using bioreporter bacteria were suggested. Recombinant *E. coli* cells carrying genes responsive to the presence of bioavailable heavy metal ions (e.g., Hg^{2+}, Cu^{2+}, Cd^{2+}, and Pb^{2+}) fused to a firefly luciferase reporter gene, were immobilized to the tip of an optical fiber and tested on 17 synthetic and 3 environmental blind samples [90]. Another Hg^{+2} monitoring system making use of the optical imaging fiber based method for Hg^{+2} monitoring, was proposed [91]. Recombinant *E. coli* bacteria, possessing the *lacZ* reporter gene, were fused to the heavy metal responsive gene promoter *zntA*, and attached to the face of an optical imaging fiber, containing a high density array of microwells. Additionally, for location identification, bacteria were introduced to the enhanced cyan fluorescent protein encoding plasmid. After 1 h incubation time, the minimum Hg^{+2} detectable concentration was established at 100 nM.

Fiber optic biosensors for Hg and As were developed by attaching alginate-immobilized recombinant luminescent Hg- and As-sensor bacteria onto optical fibers [92]. The optimized biosensors (consisting of seven layers of fiber-attached bacteria pre-grown until midlogarithmic growth phase) enabled quantification of environmentally relevant concentrations of the target analytes: 2.6 $\mu g\ L^{-1}$ of Hg(II) and 141 $\mu g\ L^{-1}$ of As(V) or 18 $\mu g\ L^{-1}$ As(III). The highest viability and sensitivity for target analyte was obtained when fiber tips were stored in $CaCl_2$ solution at −80 °C.

A whole cell fiber optic application making use of the enzymatic reaction of enzyme organophosphorus hydrolase (OPH) has been reported [51]. This method used canalization of organophosphorus nerve agents to form a chromophoric product that can absorb light at a specific wavelength. The modified bacteria are immobilized onto a bifurcated fiber optic tip and used a photomultiplier detection system to measure the light signals. Another enzymatic method makes use of inhibition of alkaline phosphatase present on the external membrane of *C. vulgaris* microalgae cells by heavy metals [93].

Detection of ionic and colloid gold with luminous *Photobacterium phosphorum* B7071 bacteria was described [94]. In this study the inhibition rate of the bioluminescent bacteria, for both gold forms, was dependent on analyte concentration and incubation time with cells.

Bioluminescent *Alcaligenes eutrophus* cells were used to monitor bioavailable heavy metal ions (Cu being chosen as a model ion) [95]. An *A. eutrophus* (AE1239) was genetically engineered by inserting a *luxCDABE* operon from *V. fischeri* under control of a copper-induced promoter. As a result, copper ions induced bioluminescence, which was proportional to the concentration of the triggering ions. The minimum detectable concentration of the sensor was established at 1 μM. Selectivity studies with Zn^{2+} showed that the sensor was selective for Cu^{2+} within the linear range.

3.1.3 Gases

Many toxic materials occur in the form of gases or suspended particles. For better human health protection, gas monitoring devices should possess characteristics such as short time of response and the ability for real-time monitoring. In some cases, for assessment of global atmospheric toxicity, biosensors methods are more effective than physicochemical ones [96]. A whole-cell biosensor was developed for the detection of gas toxicity, using recombinant bioluminescent *E. coli* harboring a *lac::luxCDABE* fusion [97]. Immobilization of the cells within LB agar was done to maintain the activity of the microorganisms, and to detect the toxicity of chemicals through direct contact with gas. The biosensor kit, in which the bioluminescent bacteria were immobilized at the far end side of a fiber optic light probe, was connected to a highly sensitive luminometer. This biosensor was able to detect the reproducible toxicity of benzene gas for different concentrations of benzene vapor. The minimum detection level was 0.2% in liquid, or nearly 48 ppm of benzene in vapor. In this case, the sensitivity of the biosensor was affected by gas diffusion into the cells. Therefore several advanced immobilization techniques were used for the improvement of sensor sensitivity [98]. Addition of glass beads during the immobilization steps enhanced the diffusion of vapors into the cell matrix and the thin layer of the matrix, allowed a more sensitive reaction. In this case the toxicity of benzene, toluene, ethylbenzene, and xylene chemicals was tested. It was found that the tested compounds toxic effect increased with chemical solubility.

3.1.4 General Toxicity

It was demonstrated that genotoxicant-responsive recombinant bioluminescent bacteria, entrapped as adlayers to an optrode sensor, may act as a potential environmental monitoring system and produce light in response to an external, inducing agent in a dose-dependent manner [29, 33]. In these studies, genetically modified *E. coli* contains a chromosomally integrated fusion of the *recA* (activates repair systems due to DNA damage) promoter region to the *Photorhabdus luminescens lux* CDABE reporter. This fiber optic method was able to detect mitomycin C as low as 25 $\mu g\ L^{-1}$ in less than 2 h. The main advantage of this method is that it does not require any solvent extraction or separation procedures before analysis, and stood as a self-contained system.

The whole cell fiber optic based method described above has thus far only been tested in static setups. So, the next step in the ultimate goal of developing a simple field monitor for relevant toxic compounds is the development of a flow through real-time fiber optic based monitoring system. Such a system for online monitoring of toxic pollutants in water has been proposed [40]. In order to monitor toxicity real-time, a field operable fiber optic photodetector device was designed in a flow-through manner previously published. These results show that the sensitivity of this system is roughly comparable to that of the static [29, 33,

99], where responses were found at 10–5,000 μg L^{-1} *p*-chlorophenol and 32–2,000 μg L^{-1} mitomycin C. The 24-h surface and tap water measurements demonstrate the ability of the device to run for such a time period. This sensor would be mostly useful as an early warning system at surface inlets for drinking water preparation, and at critical points in the distribution system.

3.1.5 Others

Although numerous, novel, microbial biosensors have been developed for measurement of various organic and inorganic compounds during the last decade, the environmental application of biosensors is still limited. Oxygen fiber optic microbiosensors were used for the quantification of available, dissolved, organic, carbon (ADOC) by microorganisms immobilized in a polyurethane hydrogel. The bacterial strain used by low substrate selectivity, and responded to mono- and disaccharides, fatty acids, and amino acids [100]. Another optical fiber biosensor for biochemical oxygen demand (BOD) is based on microorganisms coimmobilized in an osmosis matrix [101]. The sensing film for BOD measurement consists of an organically modified silicate (ORMOSIL) film, embedded with tri(4,7-diphenyl-1,10-phenanthroline) ruthenium(II) perchlorate and three kinds of seawater microorganisms immobilized on a polyvinyl alcohol sol–gel matrix. The BOD measurements were carried out in the kinetic mode inside a light proof cell at a constant temperature. Measurements were taken for 3 min followed by 10 min recovery time in 10 mg L^{-1} glucose/glutamate (GGA) BOD standard solution, and the range of determination was from 0.2 to 40 mg L^{-1} GGA. The BOD values obtained correlate well with those determined by the conventional BOD5 method for seawater samples.

The fluoresceinamine based pH fiber optic biosensor, based on calcium alginate immobilization of *Xanthobacter autotrophicus* GJ10, was used for the detection of 1,2-dichloroethane in aqueous solution [102]. In this method, enzymatic activity of the haloalkane dehalogenase *DhlA* was measured by pH change, caused by hydrolytically separating the chlorine atom from dichloroethane and generating hydrochloric acid.

Genetically modified *E. coli* cells bearing a firefly luciferase gene-fused to the TOL plasmid, immobilized on a dialysis membrane were used for detection of highly toxic benzene derivatives in a fiber optic based sensor [103]. A TOL plasmid of *P. putida* fused to the *luc* reporter system encoded enzymes for benzene degradation and thus in the presence of aromatic compounds (e.g., benzene) began to produce light. Detection limits of 5 μmol L^{-1} for *m*-xylene can be achieved by use of this biosensor, but the microbial membrane must be replaced after every use. Benzene, toluene, xylene, and their derivatives can thus be monitored.

Water toxicity detection methods, based on fluorescence fiber optic biosensors immobilized in alginate beads of *C. vulgaris* cells, have been described [104]. Colonies of immobilized microorganisms were exposed to a solution containing fluorescein diacetate (FDA), and with an intracellular, esterase, enzymatic reaction, hydrolysed FDA into a highly fluorescent compound, fluorescein. Thus, the rates of hydrolysis were affected by the overall toxicity of the tested water, and affected the overall fluorescence intensity. The working range was linear between 0.25 and 10 mg mL^{-1} FDA and the fiber optic biosensor was capable of detecting organic and metallic pollutants at levels of environmental interest.

A cell based biosensor using a single, live, bacterial, cell array platform was constructed by immobilizing live sensing bacterial cells on the distal end of a bundle of etched optical imaging fibers [105]. Imaging fibers are composed of thousands of identical, individual optical fibers, coherently bundled together. The etched fibers (1 mm in diameter and 3 cm long) contain an array of approximately 50,000 microwells, each with a diameter of 2.5 μm and a depth of 3 μm. Single cells were immobilized into each microwell, forming a high-density single cell array that allows for the simultaneous measurement of many individual cell responses. A charged coupled device (CCD) detector may be employed to monitor and spatially resolve the fluorescence signals obtained from each individual cell, allowing simultaneous monitoring of cellular responses of all the cells in the array, using reporter genes (*lac*Z, EGFP, ECFP, DsRed) or fluorescent indicators [106]. The key advantage of this system is that multiple responses from a large population of individual cells from different strains or cell lines, can be repeatedly analyzed simultaneously. Various immobilized organisms and tested compounds were reported. *E. coli* strains, containing the *lacZ* reporter gene fused to the heavy metal-responsive gene promoter *zntA*, were used to test a mercury presence [91]. NIH 3T3 mouse fibroblast cells were randomly dispersed into an optically, addressable, fiber optic, microwell array so that each microwell accommodated a single cell [107]. The cells were encoded to identify their location within the array, and to correlate changes or manipulations in the local environment to responses of specific cell types. Other applications used yeast and bacteria cell arrays that were fabricated to perform multiplexed cell assays, with resolution at the single-cell level [106]. By monitoring *lacZ* expression from individual cells of three strains of yeast – positive, negative, and wild-type – the multiplexing capability of the system was demonstrated. Fluorescence signals from individual cells were analyzed after 4 h: 33% of positive strain cells exhibited a signal increase of 100 units, while only 5% of negative strain cells showed a similar increase. No wild-type strain cells demonstrated an increase in fluorescence. This approach allows near real-time monitoring of large numbers of individual cells in an array.

3.2 Suspension Based Fiber Optic Applications

A miniature bioreactor was build as a connector between biosensing cells and toxic materials [108]. Continuous cultures were conducted in a stainless steel, miniature bioreactor with one side port a glass window for holding an optic fiber connected +to a highly sensitive luminometer. A performance evaluation measured the response to ethanol in continuous operation by using a recombinant bioluminescent *E. coli* strain. Both the ability to repeat and reproduce the induction were found to be reliable and consistent. In a continuation of this research, a two stage, whole cell, optical fiber based toxicity monitoring system for continuous analyses of aqueous samples was developed [109]. The system described was constructed from two bioreactors. The first one was run as a turbidostat, while the second, was used for the measurement of the samples. Recombinant *E. coli* containing a *RecA::luxCDABE* fusion to monitor environmental insults to DNA, with mitomycin C as a model toxicant. Pulse type exposures were used to evaluate the system's ability to reproduce and reliability. The system's ability to monitor the possible upsets or accidental discharges of toxic chemicals was also evaluated. All this data demonstrated that this two-stage minibioreactor system using recombinant bacteria containing stress promoters fused with *lux* genes is quite appropriate for continuous toxicity monitoring. This system may be developed as an early warning system in wastewater biotreatment plants, for the prevention of accidental discharge from chemical plants as well as for tracking the accidental discharge of industrial plants into nearby rivers and streams [109]. Field tests of two stage mini bioreactors showed the capabilities of this system to detect phenol and known EDCs in environmental samples [110].

Using the same technology, a portable biosensor has been developed [111] consisting of three parts: a freeze-dried biosensing strain within a vial, a small light-proof test chamber, and an optic-fiber connected between the sample chamber and a luminometer. Four genetically engineered bioluminescent bacteria – sensitive to membrane-, protein-, DNA-, and oxidative stress – were freeze-dried and used in this study. Toxicity of a sample was detected by measuring the bioluminescence, 30 min after addition to the freeze-dried strains. The results of this study demonstrated the ability of the sensor to detect the toxicity of many chemicals rapidly and reproducibly. Thus, this method may be applied as a field toxicity monitoring system due to its compactness and measures taken to ensure the strains are not released into the environment.

Another bioreactor using a bioluminescent recombinant *E. coli::luxAB* strain was developed for tributyltin on-line monitoring [112]. The structure of the biosensor features two distinct but strongly related parts. The minibioreactor allows both a stable and reproducible environment for the bacterium and an in situ contact with the toxic compound. The transducer (fiber optic) and the analytical light device allow the for real-time monitoring and analysis of the light response of bacterial cells. This fiber optic method was able to detect tributyltin as low as 0.02 μM.

Recently a novel, computerized, multisample temperature-controlled, fiber array biosensor, based on phagocyte activity was described [113]. This application allows for the simultaneous integral measurements of chemiluminescence emitted from up to six samples containing less that 0.5 μL whole blood. The optical fibers in this luminometer are used as both light guides and solid phase sample holders. This method allows monitoring of both the intra- and extracellular, phagocyte-emitted, chemiluminescent processes in the same instrument. This new technology may find use in a wide range of analytical luminescence applications in biology, biophysics, biochemistry, toxicology and clinical medicine [114].

4 Future Trends

The field of fiber optics based biosensors for environmental applications has seen great advancement in the past decade in areas such as the development of new immobilization processes and microorganisms with various genetic modifications. However, attention is needed in defining requirement and practical limits for field testing using whole cell fiber optics. Many countries are bound by strict regulations of the usage of genetically modified organisms in field testing despite the fact that they have advantages in water toxicity monitoring. Immobilization techniques need improvement and these novel methodologies will be cheaper, simpler to do, unharmful for sensor cells and will allow better diffusion of the tested compounds. Most importantly, they will help reduce leaching of sensor organisms to the tested environment. Development of genetic techniques will create better characterized bioreporter organisms with more efficient reversibility and the possibility of detection of more diverse groups of contaminants. The main disadvantage of many current biosensors is a deficiency in multianalyte detection. Thus it is likely the future will focus on the construction of multifiber arrays based on immobilized bioreporter organisms that will be capable of detecting hundreds of totally separate compounds. Finally, many biosensors require multistep protocols and sensitive reagents ill adapted for use for continually monitoring applications. However, whole cell fiber optic biosensors offer the brightest horizons with the potential to analyze samples without any additives, and will be more and more used as on-line monitoring tools. In the last decade, problems concerning pharmaceutical residues in the environment have attracted attention, stimulating research within this area. Antibiotic substances have caused special concern due to their negative influence on resistance development in bacteria. All these considerations demand new, fast and reliable techniques for the monitoring of antibacterial compounds before and after water treatment and for on-line monitoring of surface and drinking water.

References

1. Environmental Working Group (2005) A national assessment of tap water quality. EPA, California, USA
2. Belkin S (2003) Microbial whole-cell sensing systems of environmental pollutants. Curr Opin Microbiol 6(3):206–212
3. De Meulenaer B et al. (2002) Development of an enzyme-linked immunosorbent assay for bisphenol a using chicken immunoglobulins. J Agric Food Chem 50(19):5273–5282
4. Eltzov E, Kushmaro A, Marks RS (2008) Biosensors and related techniques for endocrine disruptors. In: Shaw (ed) Endocrine disrupting chemicals in food. Woodhead, Cambridge, UK
5. Sonnenschein C, Soto AM (1998) An updated review of environmental estrogen and androgen mimics and antagonists. J Steroid Biochem Mol Biol 65(1/6):143–150
6. Gu MB, Mitchell RJ, Kim BC (2004) Whole-cell-based biosensors for environmental biomonitoring and application. Adv Biochem Eng Biotechnol 87:269–305
7. Franklin NM et al. (2001) Development of an improved rapid enzyme inhibition bioassay with marine and freshwater microalgae using flow cytometry. Arch Environ Contam Toxicol 40(4):469–480

8. Riches CJ, Robinson PK, Rolph CE (1996) Effect of heavy metals on lipids from the freshwater alga Selenastrum capricornutum. Biochem Soc Trans 24(2):174S
9. Cotelle S, Ferard JF (1996) Effects of algae frozen at different temperatures on chronic assessment endpoints observed with Daphnia magna. Ecotoxicol Environ Saf 33(2):137–142
10. Orvos DR et al. (2002) Aquatic toxicity of triclosan. Environ Toxicol Chem 21(7):1338–1349
11. Chen HC et al. (1996) Neoplastic response in Japanese medaka and channel catfish exposed to N-methyl-N'-nitro-N-nitrosoguanidine. Toxicol Pathol 24(6):696–706
12. Hawkins WE et al. (1998) Carcinogenic effects of 1,2-dibromoethane (ethylene dibromide; EDB) in Japanese medaka (Oryzias latipes). Mutat Res 399(2):221–232
13. Walker WW et al. (1985) Development of aquarium fish models for environmental carcinogenesis: an intermittent-flow exposure system for volatile, hydrophobic chemicals. J Appl Toxicol 5(4):255–260
14. Ren Z et al. (2007) The early warning of aquatic organophosphorus pesticide contamination by on-line monitoring behavioral changes of Daphnia magna. Environ Monit Assess 134(1/3): 373–383
15. Borcherding J, Wolf J (2001) The influence of suspended particles on the acute toxicity of 2-chloro-4-nitro-aniline, cadmium, and pentachlorophenol on the valve movement response of the zebra mussel (Dreissena polymorpha). Arch Environ Contam Toxicol 40(4):497–504
16. De Hoogh CJ et al. (2006) HPLC-DAD and Q-TOF MS techniques identify cause of Daphnia biomonitor alarms in the River Meuse. Environ Sci Technol 40(8):2678–2685
17. Kim BC, Gu MB (2003) A bioluminescent sensor for high throughput toxicity classification. Biosens Bioelectron 18(8):1015–1021
18. Patel PD (2002) (Bio)sensors for measurement of analytes implicated in food safety: a review. Trends Anal Chem 21(2):96–115
19. Bulich A (1986) Introduction and review of microbial and biochemical toxicity screening procedures. In: Toxicity testing using microorganisms, vol. 2. CRC, Boca Raton
20. Ribo JM, Kaiser KLE (1987) Photobacterium phosphoreum toxicity bioassay. I. Test procedures and applications. Toxic Assess 2:305–323
21. Kaiser KLE, Palabrica VS (1991) Photobacterium phosphoreum toxicity data index. Water Qual Res J Can 26(3):361–431
22. Rodriguez-Mozaz S et al. (2005) Biosensors for environmental monitoring – a global perspective. Talanta 65(2):291–297
23. Marks RS et al. (2007) Handbook of biosensors and biochips. Wiley, New York
24. Collings AF, Caruso F (1997) Biosensors: recent advances. Rep Prog Phys 60(11):1397–1445
25. D'Souza SF (2001) Microbial biosensors. Biosens Bioelectron 16(6):337–353
26. Vo-Dinh T, Cullum B (2000) Biosensors and biochips: advances in biological and medical diagnostics. Fresenius J Anal Chem 366(6/7):540–551
27. Leunga A, Shankarb PM, Mutharasan R (2007) A review of fiber-optic biosensors. Sens Actuators B Chem 125:688–703
28. Marks RS et al. (1997) Chemiluminescent optical ber immunosensor for detecting cholera antitoxin. Opt Eng 36(12):3258–3264
29. Polyak B et al. (2000) Optical fiber bioluminescent whole-cell microbial biosensors to genotoxicants. Water Sci Technol 42(1/2):305–311
30. Biran I, Yu X, Walt DR (2008) Optrode-based fiber optic biosensors (bio-optrode). In: Ligler FS, Taitt CR (eds) Optical biosensors: today and tomorrow. Elsevier, Amsterdam, The Netherlands, pp 3–82
31. Premkumar JR et al. (2002) Sol-gel luminescence biosensors: encapsulation of recombinant E. coli reporters in thick silicate. Anal Chim Acta 462(1):11–23
32. Elisseeff J et al. (2000) Photoencapsulation of chondrocytes in poly(ethylene oxide)-based semi-interpenetrating networks. J Biomed Mater Res 51(2):164–171
33. Polyak B et al. (2001) Bioluminescent whole cell optical fiber sensor to genotoxicants: system optimization. Sens Actuators B Chem 74(1/3):18–26
34. Polyak B, Geresh S, Marks RS (2004) Synthesis and characterization of a biotin-alginate conjugate and its application in a biosensor construction. Biomacromolecules 5(2):389–396

35. Tombs M, Harding SE (1998) An introduction to polysaccharide biotechnology. Taylor and Francis, London
36. Gacesa P (1988) Alginates. Carbohydr Polym 8:161–182
37. Rowley JA, Madlambayan G, Mooney DJ (1999) Alginate hydrogels as synthetic extracellular matrix materials. Biomaterials 20(1):45–53
38. Yang J et al. (2002) Galactosylated alginate as a scaffold for hepatocytes entrapment. Biomaterials 23(2):471–479
39. Eiselt P, Lee KY, Mooney DJ (1999) Rigidity of two-component hydrogels prepared from alginatepolyethylene glycol diamines. Macromolecules 32:5561–5566
40. Bryant SJ, Nuttelman CR, Anseth KS (2000) Cytocompatibility of UV and visible light photoinitiating systems on cultured NIH/3T3 fibroblasts in vitro. J Biomater Sci Polym Ed 11(5):439–457
41. Eltzov E, Marks RS, Voost S, Wullings B, Heringa BM, Flow-through real time bacterial biosensor for toxic compounds in water, Sens. Actuators B: Chem, submitted, SNB-D-08-00432
42. Lu MZ et al. (2000) Cell encapsulation with alginate and alpha-phenoxycinnamylidene-acetylated poly(allylamine). Biotechnol Bioeng 70(5):479–483
43. Livage J (1997) Sol gel processes. Curr Opin Solid State Mater Sci 2:132–138
44. Yu D et al. (2005) Aqueous sol-gel encapsulation of genetically engineered Moraxella spcells for the detection of organophosphates. Biosens Bioelectron 20(7):1433–1437
45. Nassif N et al. (2003) A sol-gel matrix to preserve the viability of encapsulated bacteria. J Mater Chem 13:203–208
46. Coradin T, Nassif N, Livage J (2003) Silica-alginate composites for microencapsulation. Appl Microbiol Biotechnol 61(5/6):429–434
47. Premkumar JR et al. (2001) Antibody-based immobilization of bioluminescent bacterial sensor cells. Talanta 55(5):1029–1038
48. Bettaieb F et al. (2007) Immobilization of E. coli bacteria in three-dimensional matrices for ISFET biosensor design. Bioelectrochemistry 71(2):118–125
49. Alkorta I et al. (2006) Bioluminescent bacterial biosensors for the assessment of metal toxicity and bioavailability in soils. Rev Environ Health 21(2):139–152
50. Bitton G, Dutka BJ (1986) Introduction and review of microbial and biochemical toxicity screening procedures. In: Toxicity testing using microorganisms, vol. 2. CRC, Boca Raton, p 1–8
51. Mulchandani A et al. (1998) Biosensor for direct determination of organophosphate nerve agents using recombinant Escherichia coli with surface-expressed organophosphorus hydrolase. 1. Potentiometric microbial electrode. Anal Chem 70(19):4140–4145
52. Wu M et al. (2004) Time-resolved enzymatic determination of glucose using a fluorescent europium probe for hydrogen peroxide. Anal Bioanal Chem 380(4):619–626
53. Ramanathan S et al. (1998) Bacteria-based chemiluminescence sensing system using beta-galactosidase under the control of the ArsR regulatory protein of the ars operon. Anal Chim Acta 369(3):189–195
54. Lewis JC et al. (1998) Applications of reporter genes. Anal Chem 70(17):579a–585a
55. Burbaum JJ, Sigal NH (1997) New technologies for high-throughput screening. Curr Opin Chem Biol 1(1):72–78
56. Scheirer W (1997) Reporter gene assay applications. In: Devlin (ed) High throughput screening: the discovery of bioactive substances. Marcel Dekker, New York, pp 401–412
57. Tauriainen S et al. (1997) Recombinant luminescent bacteria for measuring bioavailable arsenite and antimonite. Appl Environ Microbiol 63(11):4456–4461
58. Meighen EA (1988) Enzymes and genes from the lux operon of bioluminescent. Annu Rev Microbiol 42:151–176
59. McElroy WD, Seliger HH (1963) The chemistry of light emission. In: Advances in enzymology. Wiley, New York, pp 119–162
60. Meighen EA, Dunlap PV (1993) Physiological, biochemical and genetic-control of bacterial bioluminescence. Adv Microb Physiol 34:1–67
61. Meighen EA, Szittner RB (1992) Multiple repetitive elements and organization of the lux operons of luminescent terrestrial bacteria. J Bacteriol 174(16):5371–5381

62. Belkin S et al. (1997) A panel of stress-responsive luminous bacteria for the detection of selected classes of toxicants. Water Res 31(12):3009–3016
63. Davidov Y et al. (2000) Improved bacterial SOS promoter:: lux fusions for genotoxicity detection. Mutat Res Genet Toxicol Environ Mutagen 466(1):97–107
64. Vollmer AC et al. (1997) Detection of DNA damage by use of Escherichia coli carrying recA'-lux, uvrA'-lux, or alkA'-lux reporter plasmids. Appl Environ Microbiol 63(7):2566–2571
65. Vandyk TK et al. (1994) Rapid and sensitive pollutant detection by induction of heat-shock gene-bioluminescence gene fusions. Appl Environ Microbiol 60(5):1414–1420
66. Michael B et al. (2001) SdiA of Salmonella enterica is a LuxR homolog that detects mixed microbial communities. J Bacteriol 183(19):5733–5742
67. Goh EB et al. (2002) Transcriptional modulation of bacterial gene expression by subinhibitory concentrations of antibiotics. Proc Natl Acad Sci U S A 99(26):17025–17030
68. Eltzov E et al. (2008) Detection of sub-inhibitory antibiotic concentrations via luminescent sensing bacteria and prediction of their mode of action. Sens Actuators B Chem 129(2):685–692
69. Dukan S et al. (1996) Hypochlorous acid activates the heat shock and soxRS systems of Escherichia coli. Appl Environ Microbiol 62(11):4003–4008
70. Pedahzur R, Shuval HI, Ulitzur S (1997) Silver and hydrogen peroxide as potential drinking water disinfectants: their bactericidal effects and possible modes of action. Water Sci Technol 35(11/12):87–93
71. Van Dyk TK et al. (1998) Constricted flux through the branched-chain amino acid biosynthetic enzyme acetolactate synthase triggers elevated expression of genes regulated by rpoS and internal acidification. J Bacteriol 180(4):785–792
72. Ben-Israel O, Ben-Israel H, Ulitzur S (1998) Identification and quantification of toxic chemicals by use of Escherichia coli carrying lux genes fused to stress promoters. Appl Environ Microbiol 64(11):4346–4352
73. Daunert S et al. (2000) Genetically engineered whole-cell sensing systems: coupling biological recognition with reporter genes. Chem Rev 100(7):2705–2738
74. Ashley JTF, Baker JE (1999) Hydrophobic organic contaminants in surficial sediments of Baltimore Harbor: inventories and sources. Environ Toxicol Chem 18(5):838–849
75. Arcaro KF et al. (1999) Antiestrogenicity of environmental polycyclic aromatic hydrocarbons in human breast cancer cells. Toxicology 133(2/3):115–127
76. Vinggaard AM, Hnida C, Larsen JC (2000) Environmental polycyclic aromatic hydrocarbons affect androgen receptor activation in vitro. Toxicology 145(2/3):173–183
77. Gu MB, Chang ST (2001) Soil biosensor for the detection of PAH toxicity using an immobilized recombinant bacterium and a biosurfactant. Biosens Bioelectron 16(9/12):667–674
78. Heitzer A et al. (1994) Optical biosensor for environmental on-line monitoring of naphthalene and salicylate bioavailability with an immobilized bioluminescent catabolic reporter bacterium. Appl Environ Microbiol 60(5):1487–1494
79. Ripp S et al. (2000) Controlled field release of a bioluminescent genetically engineered microorganism for bioremediation process monitoring and control. Environ Sci Technol 34(5):846–853
80. Yolcubal I et al. (2000) Fiber optic detection of in situ lux reporter gene activity in porous media: system design and performance. Anal Chim Acta 422(2):121–130
81. Fine T et al. (2006) Luminescent yeast cells entrapped in hydrogels for estrogenic endocrine disrupting chemical biodetection. Biosens Bioelectron 21(12):2263–2269
82. Horry H et al. (2007) Technological conception of an optical biosensor with a disposable card for use with bioluminescent bacteria. Sens Actuators B Chem 122(2):527–534
83. Hotchkiss AK et al. (2002) Androgens and environmental antiandrogens affect reproductive development and play behavior in the Sprague-Dawley rat. Environ Health Perspect 110(Suppl 3):435–439
84. Moore RW et al. (2001) Abnormalities of sexual development in male rats with in utero and lactational exposure to the antiandrogenic plasticizer di(2-ethylhexyl) phthalate. Environ Health Perspect 109(3):229–237

85. Kumar J, Jha SK, D'Souza SF (2006) Optical microbial biosensor for detection of methyl parathion pesticide using Flavobacterium sp whole cells adsorbed on glass fiber filters as disposable biocomponent. Biosens Bioelectron 21(11):2100–2105
86. Gammon DW et al. (2005) A risk assessment of atrazine use in California: human health and ecological aspects. Pest Manage Sci 61(4):331–355
87. Frense D, Muller A, Beckmann D (1998) Detection of environmental pollutants using optical biosensor with immobilized algae cells. Sens Actuators B Chem 51(1/3):256–260
88. Vedrine C et al. (2003) Optical whole-cell biosensor using Chlorella vulgaris designed for monitoring herbicides. Biosens Bioelectron 18(4):457–463
89. Naessens M, Leclerc JC, Tran-Minh C (2000) Fiber optic biosensor using Chlorella vulgaris for determination of toxic compounds. Ecotoxicol Environ Saf 46(2):181–185
90. Hakkila K et al. (2004) Detection of bioavailable heavy metals in EILATox-Oregon samples using whole-cell luminescent bacterial sensors in suspension or immobilized onto fibre-optic tips. J Appl Toxicol 24(5):333–342
91. Biran I et al. (2003) Optical imaging fiber-based live bacterial cell array biosensor. Anal Biochem 315(1):106–113
92. Ivask A et al. (2007) Fibre-optic bacterial biosensors and their application for the analysis of bioavailable Hg and As in soils and sediments from Aznalcollar mining area in Spain. Biosens Bioelectron 22(7):1396–1402
93. Durrieu C, Tran-Minh C (2002) Optical algal biosensor using alkaline phosphatase for determination of heavy metals. Ecotoxicol Environ Saf 51(3):206–209
94. Gruzina TG et al. (2005) Luminescent test based on photobacterium phosphorum b7071 for ionic and colloid gold determination in water. Khimiya i Tekhnologiya Vody 27: 200–208
95. Leth S et al. (2002) Engineered bacteria based biosensors for monitoring bioavailable heavy metals. Electroanalysis 14(1):35–42
96. Dennison MJ, Hall JM, Turner APF (1995) Gas-phase microbiosensor for monitoring phenol vapor at Ppb levels. Anal Chem 67(21):3922–3927
97. Gil GC et al. (2000) A biosensor for the detection of gas toxicity using a recombinant bioluminescent bacterium. Biosens Bioelectron 15(1/2):23–30
98. Gil GC, Kim YJ, Gu MB (2002) Enhancement in the sensitivity of a gas biosensor by using an advanced immobilization of a recombinant bioluminescent bacterium. Biosens Bioelectron 17(5):427–432
99. Pedahzur R et al. (2004) Water toxicity detection by a panel of stress-responsive luminescent bacteria. J Appl Toxicol 24(5):343–348
100. Koster M, Gliesche CG, Wardenga R (2006) Microbiosensors for measurement of microbially available dissolved organic carbon: sensor characteristics and preliminary environmental application. Appl Environ Microbiol 72(11):7063–7073
101. Lin L et al. (2006) Novel BOD optical fiber biosensor based on co-immobilized microorganisms in ormosils matrix. Biosens Bioelectron 21(9):1703–1709
102. Campbell DW, Muller C, Reardon KF (2006) Development of a fiber optic enzymatic biosensor for 1,2-dichloroethane. Biotechnol Lett 28(12):883–887
103. Ikariyama Y et al. (1997) Fiber-optic-based biomonitoring of benzene derivatives by recombinant E-coli bearing luciferase gene-fused TOL-plasmid immobilized on the fiber optic end. Anal Chem 69(13):2600–2605
104. Merchant D et al. (1998) Optical fibre fluorescence and toxicity sensor. Sens Actuators B Chem 48(1/3):476–484
105. Biran I et al. (2003) Optical imaging fiber-based live bacterial cell array biosensor. Anal Biochem 315(1):106–113
106. Biran I, Walt DR (2002) Optical imaging fiber-based single live cell arrays: a high-density cell assay platform. Anal Chem 74(13):3046–3054
107. Taylor LC, Walt DR (2000) Application of high-density optical microwell arrays in a live-cell biosensing system. Anal Biochem 278(2):132–142

108. Gu MB et al. (1996) A miniature bioreactor for sensing toxicity using recombinant bioluminescent Escherichia coli cells. Biotechnol Prog 12(3):393–397
109. Gu MB, Gil GC, Kim JH (1999) A two-stage minibioreactor system for continuous toxicity monitoring. Biosens Bioelectron 14(4):355–361
110. Gu MB et al. (2001) The continuous monitoring of field water samples with a novel multichannel two-stage mini-bioreactor system. Environ Monit Assess 70(1/2):71–81
111. Choi SH, Gu MB (2002) A portable toxicity biosensor using freeze-dried recombinant bioluminescent bacteria. Biosens Bioelectron 17(5):433–440
112. Thouand G et al. (2003) Development of a biosensor for on-line detection of tributyltin with a recombinant bioluminescent Escherichia coli strain. Appl Microbiol Biotechnol 62(2/3):218–225
113. Magrisso M et al. (2006) Fiber-optic biosensor to assess circulating phagocyte activity by chemiluminescence. Biosens Bioelectron 21(7):1210–1218
114. Prilutsky D et al. (2008) Dynamic component chemiluminescent sensor for assessing circulating polymorphonuclear leukocyte activity of peritoneal dialysis patients. Anal Chem 80(13):5131–5138

Adv Biochem Engin/Biotechnol (2010) 117: 155–178
DOI:10.1007/10_2009_5

Published online: 21 May 2009

Electronic Interfacing with Living Cells

James T. Fleming

Abstract The direct interfacing of living cells with inorganic electronic materials, components or systems has led to the development of two broad categories of devices that can (1) transduce biochemical signals generated by biological components into electrical signals and (2) transduce electronically generated signals into biochemical signals. The first category of devices permits the monitoring of living cells, the second, enables control of cellular processes. This review will survey this exciting area with emphasis on the fundamental issues and obstacles faced by researchers. Devices and applications that use both prokaryotic (microbial) and eukaryotic (mammalian) cells will be covered. Individual devices described include microbial biofuel cells that produce electricity, bioelectrical reactors that enable electronic control of cellular metabolism, living cell biosensors for the detection of chemicals and devices that permit monitoring and control of mammalian physiology.

Keywords Bioelectrical reactors • Bioelectronics • Biofuel cells • Biosensors • Brain-machine interface • Cellular FETs

Contents

J.T. Fleming
University of Tennessee, Center for Environmental Biotechnology, 676 Dabney Hall, Knoxville, 37996, TN USA
e-mail: jtf@utk.edu

1 Bioelectronics and Live Cell Interfacing

The fields of electrochemistry and electrophysiology, which form the basis of modern research dealing with the interactions between electricity and biological systems, originated with Galvani's experiments innervating muscle tissue with static electricity in the eighteenth century. This was a crucial experiment in the historical development of electrical theory in that it was in response to Galvani's results that his contemporary, Volta, invented the early battery, which quickly led to the modern understanding of the physical basis of electrical charge transfer.

A new area of research is now emerging from a union of the fields of electronics, materials sciences and biology, termed "bioelectronics" that promises revolutionary advances in medicine, agriculture, industrial processes and military applications. The domain of bioelectronics is concerned with the interfacing of naturally occurring or synthetic biological materials with inorganic electronic materials, components or systems. This has led to the development of two broad categories of devices that can (1) transduce biochemical signals generated by biological components into electrical signals and (2) transduce electronically generated signals into biochemical signals. The first category of devices permits the monitoring of living cells, the second, enables control of cellular processes.

This review will focus on one particular area within the field of bioelectronics dealing with the electronic interfacing of living cells. Topics that will be considered here are methods or devices that enable control or monitoring of living cells by direct electronic interface with an emphasis on the fundamental principles involved. The term "direct" is taken to mean cells in physical contact with electrodes or chips the interface of which has an electrochemical basis. Examples will be drawn from the scientific literature describing the use of both prokaryotic (microorganisms) and eukaryotic (mammalian) cells to illustrate these approaches. For the purposes of this chapter, approaches or devices that use individual cellular components (e.g., purified enzymes) will not be considered. Likewise, photonic or hybrid photonic/ bioelectronic devices, will not be considered because they do not meet the direct interface criteria. A systematic evaluation of electrode materials, while certainly pertinent to the topics presented, will not be discussed because it is simply beyond the scope of this short review.

2 Fundamental Considerations Involved in the Electronic Interfacing of Living Cells

There are a number of principles and considerations that are common to all electronic/ living cell interfacing approaches. These include consideration of the electrochemical reactions that may occur at the electrodes, possible electroporation effects, electron transfer mechanisms, the pros and cons of using living cells and considerations involved in the immobilization of cells on devices.

2.1 Electrochemical Reactions at the Electrode surface

When two electrodes are placed in an electrolyte and an electric current is applied, one of the electrodes is referred to as the "working electrode" and the other as the "counter electrode." The working electrode is the one where the processes being studied occur; the counter electrode completes the circuit to the working electrode. A third electrode may be used, referred to as the "reference electrode" as a reference for electrical potential measurements.

Charge transfer may occur by one of two mechanisms – Faradaic or nonFaradaic processes. Faradaic reactions are characterized by a flow of electrons between the electrode and the electrolyte resulting in reduction or oxidation of species in the electrolyte; nonFaradiac reactions are characterized as having no flow of electrons. Charge transfer in the later case occurs by way of the redistribution of charged species in the electrolyte [1].

Any device design that uses Faradaic processes must consider the biological effects of a series of reactions that may occur under certain circumstances. These reactions are dependent on the ions in solution and the potential difference between the electrodes [2]. If the potential difference between working electrode and the counter electrode is greater than 1.23 V, electrolysis of water will occur leading to production of O_2 and H^+ at the anode and H_2 at the cathode:

$$O_{2(g)} + 4H^+ + 4e^- \rightleftharpoons 2H_2O, \quad E = 1.229\,V.$$

At a potential of greater than 2 V, production of OH^- occurs:

$$2H_2O + 2e^- \rightleftharpoons 2H_{2(g)} + 2OH^-, \quad E = -0.828\,V.$$

An increase in $[OH^-]$ can lead to a change in the pH of the solution.

If chlorine ions are present, as they are in physiological saline solutions, production of chlorine gas may also occur:

$$Cl_{2(g)} + 2e^- \rightleftharpoons 2Cl^-, \quad E = 1.358\,V.$$

Hydrogen peroxide may also be formed:

$$O_{2(g)} + 2H + 2e^- \text{ € } H_2O_2, \quad E = 0.695\,V.$$

If carbon electrodes are used, the following reaction may occur:

$$CO_2 + +4H + +4e^- \text{ € } C + 2H_2O, \quad E = -0.213V.$$

Faradaic reactions can also have degradative effects on electrodes. To illustrate consider the corrosion of platinum in the presence of chloride ion:

$$Pt + Cl^- \text{ € } [PtCl_4]^{-2} + 2e^-$$

or the dissolution of an iron electrode:

$$Fe \text{ € } Fe^{+2} + 2e^-, \quad E = 0.440\,V.$$

Faradaic reactions may be reversible or irreversible depending on the relative rates of two processes: electron transfer at the interface and mass transport. A reaction with fast electron transfer and slow mass transport is reversible. In this case, the products formed do not diffuse away from the electrode quickly and, therefore, if the current charge is reversed, some of the formed product may be transformed back into reactants. If the relative rates of the two processes are reversed resulting in a low electron transfer rate and fast mass transport, the products will move away from the electrode before the reaction can be reversed and the reaction will be effectively irreversible [1].

2.2 Electroporation

The issue of possible electroporation damage to interfaced cells has been raised recently [3]. Electroporation is a technique for delivering molecules into impermeable living cells by means of an electric field ranging from 0.3 kV cm^{-1} to 12 kV cm^{-1} with a pulse duration of 0.5 μs to 50 ms [4].

The membrane voltage, V_m of a spherical cell membrane in a homogeneous electric field for duration t can be derived from the equation

$$V_m = 1.5 r_c \; E \cos\alpha [1 - \exp(-t / \tau_m)],$$

where E is the electric field strength, r_c is the radius of the cell, α is the angle in relation to the direction of the electric field and

$$\tau_m = r_c C_m ((R_{int} + R_{ext}) / 2)$$

is the membrane relaxation time. C_m is the membrane capacitance, and R_{int} and R_{ext} are the specific resistivities of the intracellular and extracellular media [5]. When the membrane potential reaches a critical value of 0.2-1.5 V for mammalian cells, membrane breakdown will occur and pores will form in the membrane [5]. In bulk electroporation, populations of cells are exposed to homogeneous fields of ~1 kV cm^{-2} using macroelectrodes separated by millimeters [6]. Recently, techniques have been developed to electroporate single cells on silicon chips. For chip-based single cell methods lower field strengths of only ~1 V cm^{-2} are required [6]. It is, therefore, certainly possible that electroporation effects may be incurred with cells fixed to microelectrodes or other silicon-based devices. As will be noted below, several groups have designed systems that that stimulate cells using capacitive currents, one advantage of which is that possible electroporation effects are eliminated (Sect. 3.4.2). However, while the issue of possible electroporation effects is certainly legitimate, there is presently no published experimental evidence to indicate that it occurs with chip-based Faradaic devices.

2.3 Electron Transfer Mechanisms

The transfer of electrons between a living cell and an electrode may occur directly or by way of a molecular shuttling intermediate or by the electrolysis of water [2] (Fig. 1). For direct electron transport to occur the electrode must contact the outer membrane of the cell [7]. If this proximity condition has been met, electrons can hop to the electrode from membrane bound enzymes. Concerning mediated transfer, many types of molecules, generally with conjugated ring structures, can function as electron transfer intermediates which, after being oxidized or reduced directly at the electrodes are, in turn, oxidized or reduced by the cell.

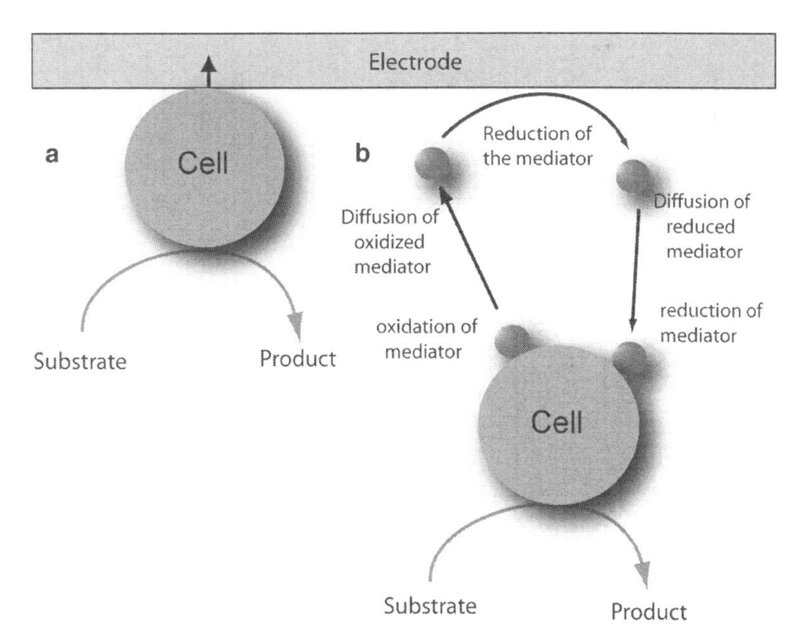

Fig. 1 Electron transfer from electrode to cell. **a** Direct transfer. **b** Mediated transfer

The precise mechanisms for electron transfer to living cells are not well understood. Details relating to electron transfer will be considered in the application sections below when pertinent. Further understanding of electron transfer mechanisms will be necessary to improve the efficiency and fully utilize the bioelectronic processes and devices described below.

2.4 *Advantages and Disadvantages of Using Living Cells*

The use of living cells as sensor systems has several advantages and disadvantages compared to the use of single component-based bioelectronic devices, such as immobilized enzymes. Living cells provide more information in response to a stimulus compared to enzyme-based biosensors. Because cells respond to a stimulus by many parallel pathways, assessment of multiple pathway parameters has the potential to yield system redundancy and therefore higher confidence in the interpretation of the data. From the standpoint of using living cells as biosensors, the response to a stimulus is immediately and directly biologically relevant. Thus, with the use of whole cell sensors, functional biological information is obtained directly. This is in contrast to the relevance of analytical information which is used to infer functional significance based on consensual criteria or comparison with the values obtained from the scientific literature. The use of whole cells potentially confers greater stability to a sensing system, compared to the use of isolated cellular components because, in living cells, the cellular components undergo continual repair and replacement [8].

A limitation to the use of whole cells as biosensors is the slow diffusion of chemicals across the cell membrane which results in a slower response time compared with isolated cellular component sensors [9]. Cells can be made more permeable by treatment with detergents or organic chemicals but these treatments make the cells less viable [9]. Another drawback to the use of whole cells is their low specificity compared with isolated component sensors.

2.5 *Immobilization of Cells*

For efficient electron transfer between cells and the transducer, cells must be immobilized in close proximity to the transducer element. The approaches used differ markedly for prokaryotes and eukaryotes primarily because of the physiological differences between the two types of cells. It is therefore convenient to discuss immobilization approaches for these two categories of cells separately.

While numerous immobilization methods have been developed for prokaryotic cells, they can be categorized into three types: (1) adsorption, (2) chemical methods and (3) entrapment [9]. The actual stability of immobilized cells is dependent on the cell type and the particular application. For example, a comparison of bacterium-based methane biosensors indicated that cells were stable on the order of months [10].

Adsorption involves simply incubating a cell suspension with the electrode or transducer element. The mechanism of attachment is due weak interactions by way of ionic, polar, hydrogen bonding or hydrophobic interactions [11]. The main advantage of this method is that cell viability is not compromised by the use of harsh chemicals; the downside is that cells are easily desorbed giving a limited lifetime to the system. In addition, desorption may be enhanced by changes in pH, ionic strength and temperature [12].

Entrapment involves the retention of cells in proximity of the electrode by means of a permeable physical barrier such as filters and polymer encasement. This approach uses mild conditions and is therefore advantageous for maintaining cell viability. A common procedure is the use of dialysis membranes to retain cells. More sophisticated methods include encapsulation in soft gels including agarose, polyacrylamide, alginates, collagen, cellulose triacetate, silicone rubber and sol-gel silicates [8, 12].

Many prokaryotes naturally form biofilms that effectively entrap and protect cells. If particular nutrients become limiting during growth, cells adapt by changing their physiological state from exponential to stationary phase. During this transition, polysaccharides are exported which function as a polymer encasing material within which the exuding cells are suspended. This exopolysaccharide matrix is permeable to nutrients and enables survival in harsh environments. Biofilm formation plays a key role in the efficiency microbial fuel cells as will be discussed in Sect. 3.1.

Chemical methods include (1) covalent binding between functional groups on cells and the transducer and (2) cross-linking. The difficulty with covalent chemical methods is that the harsh chemicals used tend to damage cell membranes. To overcome this problem groups have immobilizing cells by way of covalently bound antibodies to cell surface proteins [13]. Glyceraldehyde-based cross linking to protein supports has also been described but these methods affected cell viability [8]. A recent paper describes a mild method for the immobilization of *Pseudomonas fluorescens* to the surface of a carbon-nanotube epoxy composite electrode. Cells were first encased in gelatin which was then cross-linked with gluteraldehyde [14]. Another recent paper describes entrapping *P. aeruginosa* and *Klebsiella sp.* cells together in polyvinyl alcohol for monitoring methane. The stability of the immobilized cells beads were assayed at 3 day intervals and it was found that cells lost 50% of their activity over a period of 1 month [10]. The requirement to have cells immobilized and simultaneously maintain their viability is one of the main stumbling blocks that currently limit more widespread implementation of this technology.

While eukaryotic cells obtained from tissues can be grown in suspension, they tend to adhere naturally to surfaces. This is due to the role played by extracellular adhesion proteins, such as fibronectin, that play an important role in cell adhesion, morphology and migration. These cell-surface processes are required for tissue repair and contribute to survival, growth and differentiation [15]. Cells can be made to adhere to transducer surfaces by modifying the substrate material with adhesive proteins or peptides that mimic the cohesive tissue matrix environment. Commonly used extracellular adhesion proteins include collagen, laminin, fibronectin and poly-L-lysine [16]. Eukaryotic cells can also be encapsulated into hydrogels to form perfused 3D cultures [16].

2.6 *Maintaining Cell Viability*

The main disadvantage to using living cells as sensors is keeping them alive under experimental conditions that may be harsh. The problems associated with immobilizing cells on a device are related to this issue because techniques used to fix cells to surfaces (e.g., crosslinking) can damage cells.

Embedding electrodes in tissues (i.e., brain) for long term recording requires that the issue of biocompatibility be addressed. It has been the experience of researchers that implanted electrodes lose their functionality over time. For example, one study using rhesus monkeys reported a 40% drop in the number of functional brain implanted electrodes over a period of 18 months [17]. This loss of functionality has been ascribed to the immune response mounted against the implanted electrode by the surrounding cells. In neural tissue, glial cells, including astrocytes and microglia, are the cells primarily involved in the immune response to device implants. When activated by injury these cells demonstrate increased migration, proliferation, and hypertrophy [18]. The mechanical trauma of the implant causes damage to blood vessels which release erythrocytes, platelets, clotting factors and the compliment cascade that causes macrophage recruitment and initiates tissue repair [18]. This initial acute response subsides after several weeks but is replaced with a chronic inflammatory response that is characterized by the presence of activated astrocytes and microglial cells and the formation of an encapsulation layer around the electrode called a "glial scar." Currently, it is thought that this encapsulation layer functions to insulate the electrode from the nerve cells increasing the impedance [18]. In response to this issue electrodes have been modified in an attempt to reduce or eliminate the immune response effect; examples of this approach are described in Sect. 3.4.2.

2.7 *Electrical Propagation in Electrogenic Cells*

Electrogenic cells, including nerve, endocrine and muscle cells, have ion channels which are electrically gated such that a change in potential results in ion transport. Ion channels are transmembrane proteins that function to control the transport of ions across the membrane. The relationship between the electrical gradient and the ionic gradient is described by the Nernst equation [19]:

$$V = RT/nF \ln(C_e / C_i),$$

where the V is the equilibrium potential = $V_{\text{cytoplasm}} - V_{\text{extracellular}}$, C_e is the extracellular ion concentration, C_i is the intracellular ion concentration, R is the gas constant, F is Faraday's constant, T is the temperature in degrees Kelvin, and n is the charge on the ion (1.6×10^4 coulombs V^{-1} mol^{-1}).

Because the equilibrium potentials of the three ions most important to action potentials, sodium (Na^{+1}), potassium (K^{+1}) and chloride (Cl^{-1}) differ, the Nernst equation may be modified to take into account the relative permeabilities of each ion [20]:

$$V = RT / F \ln \frac{PK[K^{\pm}]_{out} + PNa[Na^{\pm}]_{out} + PCl[Cl^{-}]_{in}}{PK[K+]_{in} + PNa[Na^{+}]_{in} + PCl[Cl^{-}]_{out}}.$$

In the resting state the permeability of the membrane to *K* is low such that the *V* is close to the equilibrium potential of *K* ~−75 mV. If the membrane is depolarized the inward flow of Na increases more than the outward flow of *K* such that E_m approximates the equilibrium potential of Na ~ + 55 mV which initiates the action potential peak. As the potential approaches +55 mV, the Na channels close and the membrane voltage drops. This action potential propagates as a wave along the electrogenic cell membrane which, if strong enough, may pass to the membranes of other adjacent cells.

For nerve cell stimulation, current is generally supplied in biphasic pulses, each of which has both a cathodic and anodic phase such that the overall net charge is zero. The cathodic current reduces with the direction of flow from the electrode to the cell; the anodic current oxidizes with the direction of flow from the cell to the electrode [1]. The intention of charge-balanced stimulation is to reduce damage to the electrodes and tissue by preventing the irreversible reduction and oxidation reactions outlined in Sect. 2.1 [1, 21]. Considering the high voltages required to overcome impedance during stimulation, it is unlikely that electrode or tissue damage can be completely prevented [22].

3 Approaches to the Electronic Interfacing of Living Cells

The interfacing of living cells with electronic devices has been accomplished in a number of different fields for many different purposes. It is useful, therefore, to organize a discussion of live cell interfacing around the prominent applications. Some of these applications include microbial fuel cells, biochemical reactors, whole cell biosensors and the study of cellular physiology.

3.1 Microbial Fuel Cells

Microbial fuel cells (MFC) extract electric current from the metabolic processes of microorganisms. This process was first reported in 1911 where a voltage difference was observed when electrodes were placed in cultures of *E. coli* [23]. Over the past decade, work in this area has dramatically expanded as the necessity for developing alternatives to fossil fuel based energy technologies has been recognized.

Evidence for naturally occurring extracellular electron transfer was first demonstrated in bacteria where electrons derived from the oxidation of organic matter are transferred to minerals such as iron and manganese oxides, leading to their reduction [24]. For this reason, metal oxide reducers, such as *Geobacteraceae* and *Shewanella* species, are commonly used for MFCs. MFCs are composed of two electrodes, the anode and cathode, separated by a semipermeable membrane with the microbes in contact with the anode. When a suitable substrate is added to the system, it is oxidized at the anode and the released electrons travel to the cathode by way the circuit connection. At the cathode, O_2 functions as the terminal electron acceptor and water is produced (Fig. 2). This process in fuel cells is, therefore, analogous to the process that occurs in nature with the solid electrode substituting for the metal oxide [7]. The maximum power output of MFCs reported to date is ~500 W m^{-3} which approaches the electrical output of anaerobic digesters (1,000 W m^{-3}) [25]. The first applications for MFC electricity generation will be most likely be specialized tasks such as powering sensors in remote locations [26].

Both mediated and direct electron transfer MFC designs have been described. In a design illustrating the mediated approach, thionine was added to a *Proteus vulgaris* suspension compartmentalized around the anode. Using glucose as fuel

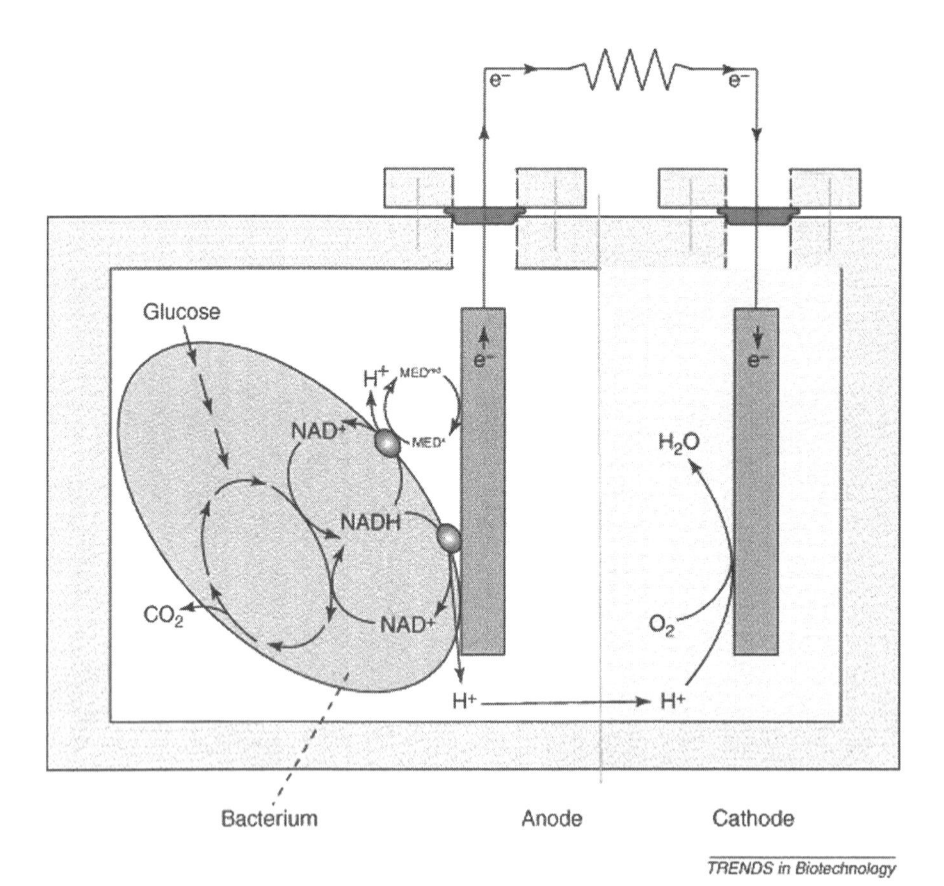

Fig. 2 Microbial fuel cell. Substrate (glucose) is metabolized by bacteria which transfer the electrons to the anode. Both direct and mediated (med) mechanisms are illustrated. Used from Rabaey et al. [24] with permission

yielded an efficiency of 89% and a power output of 360 μW cm^{-2} [27]. In another mediated design a miniature fuel cell was described using *Shewanella oneidensis*. The addition of mediators lactate and ferricyanide resulted in power increases of 30–100% to give maximum power output of 3 Wm^{-2}or 500 Wm^{-3} [28].

More recently, direct electron transfer between microorganisms and electrodes without the use of mediators has been demonstrated in MFCs. The evidence for direct anodic electron transfer is extensive [29]. Evidence for direct transfer by this mechanism was proven using *S. putrefaciens*. During anaerobic growth accumulation of a "C" type cytochrome occurs in the membrane of *S. putrefaciens*. The *omcB* gene, which codes for the C cytochrome, was inactivated by genetic techniques, resulting in a 45% reduction in ferric oxide reduction, thus proving its involvement in direct electron transfer [30]. Additional support for a role for C cytochromes in electron transfer came from a recent report that used surface enhanced infrared absorption spectroscopy and subtractively normalized interfacial Fourier transform infrared spectroscopy to detect turnover of oxidized/reduced states in C cytochromes [31].

Additional evidence for direct electron transfer came from MFCs using *Geobacter* species. These systems do not produce a detectable electron shuttle yet function effectively in high current density MFCs [32]. In this system, formation of a thick (75 μm) structured biofilm is necessary for electron transfer to the anode. The mechanism of electron transfer is biofilms is not well understood [33].

Only one instance of direct cathodic electron transfer to a bacterial cell exists in the literature where *G. metallireducens* and *G. sulfurreducens* were demonstrated to use electrons directly from an electrode to reduce nitrate and fumerate respectively [34]. While the mechanism for direct transfer has not been conclusively demonstrated in this system, it has been conjectured that cathodic electrons are passed directly to the nitrate reductase resulting in the reduction of nitrate [2]. Also conjectured are structures on the surface of microbes, "microbial nanowires," that function to conduct extracellular electron flow [33] (Fig. 3). To date, electron transfer through such filamentous structures has not been directly demonstrated [33].

3.2 Bioelectrical Reactors

Bioelectrical reactors (BER) share many design characteristics with MFCs. However, in the case of BERs, electrical current is supplied to microorganisms with the intention of manipulating metabolic processes. While this scheme, at the outset, would appear to be very modern, the underpinnings for the process can be dated to the 1930s when the mechanistic basis of cellular metabolism was demonstrated to be electrochemical. Metabolism was shown to be a process whereby a chemical substrate serves as an electron donor and the shuttling of those donated electrons through the glycolytic and electron transport chains leads to a series of linked oxidation/reduction reactions that energetically power anabolic cellular processes. It was only later, in the 1950s, that scientists actually attempted to use electricity to

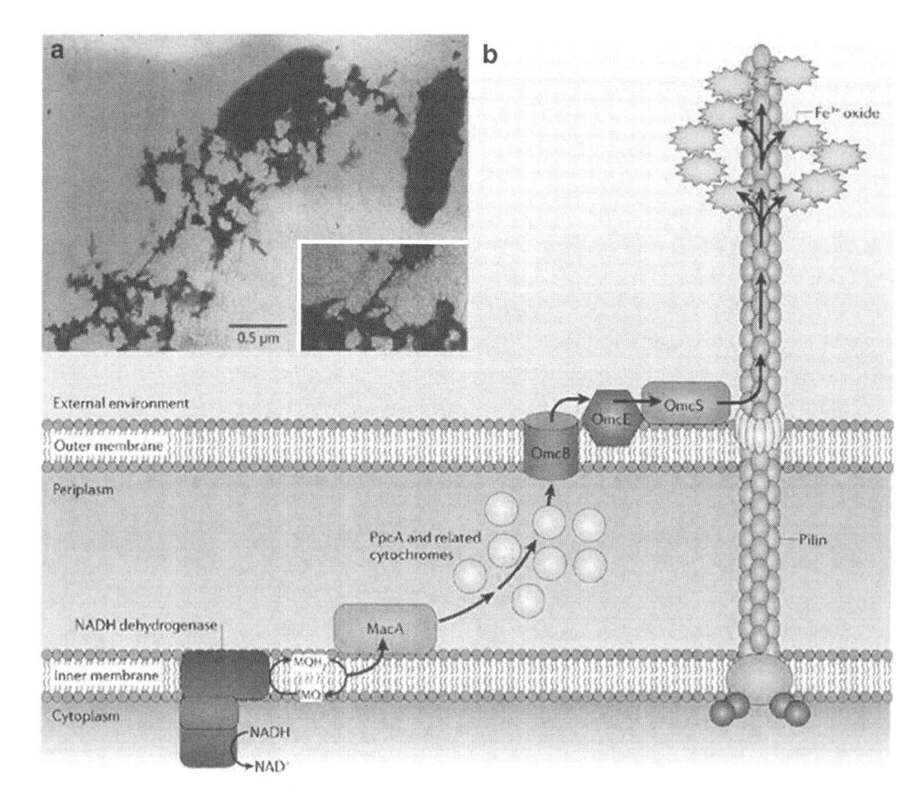

Fig. 3 A mechanism for extracellular electron transfer by *Geobacter sulfurreducens*. **a** Transmission electron micrograph showing the association of Fe^{+3} oxide (*arrows*) with pilin expressed by *Geobacter sulfurreducens*. **b** Potential route for electron transfer to Fe^{+3} oxides by *Geobacter sulfurreducens*. MacA, PpcA, OmcB, OmcE and OmcS are C-type cytochromes. Used with permission from Lovley [33]

manipulate cellular metabolism. One of these early studies used water electrolysis as a means of supplying oxygen to submerged cultures of *P. fluorescens*. Electrolysis driven growth rates equivalent to aerated cultures were obtained but necessitated that the voltage applied be below the potential for chlorine gas generation to prevent toxicity [35].

Bioelectrical reactors have been developed for the culturing of organisms, production of specific metabolites, and detoxification of hazardous chemicals. BER designs use both direct interfacing of an electrode with a cellular electron transport component and indirect stimulation involving the transfer of electrons from an electrode to the electron transport system via a soluble mediator [2]. Two alternate approaches are used for bioelectric reactors: (1) direct or indirect cathodic reduction of an electron transport component which passes the electrons to a terminal reductase which, in turn, reduces an oxidized substrate and (2) generation of a continuous

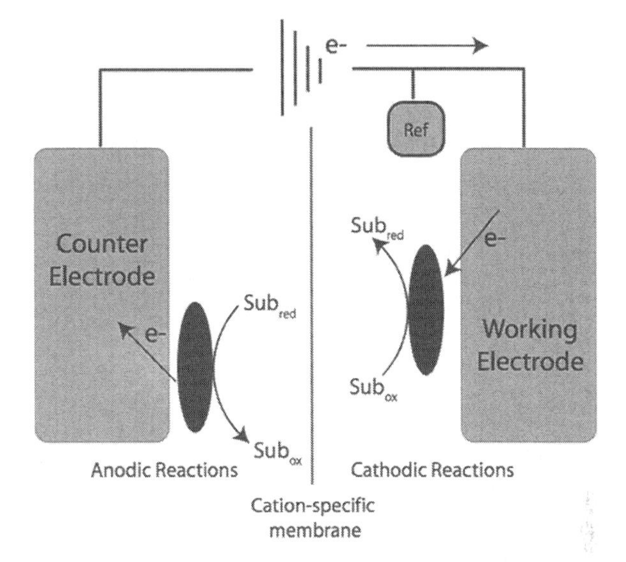

Fig. 4 Bioelectrical reactor. BERs can stimulate microbial metabolism by acting as cathodic electron sources or anodic electron sinks. In each case, reduction or oxidation, respectively, of a substrate is coupled to the electrical stimulus by the microorganism. Used by permission from Thrash [2]

supply of electron acceptors by direct or indirect anodic oxidation which are used to oxidize a reduced substrate [2](Fig. 4).

This approach has been used to modify the *Actinobacillus succinogenes* fermentation products of glucose by using electron shuttles, such as neutral red, in conjunction with cathodic reduction. Neutral red was continuously reduced with a potential of −1.5 V at 0.3–1.0 mA which permitted anaerobic growth on fumerate to produce succinate. Electrically reduced neutral red thus replaced H_2 as the sole electron donor for succinate production. By this method a 20–40% increase in succinate was obtained compared with nonelectricity driven fermentation [36].

BERs have also been used to engineer biotransformation processes. Reduction of 6-bromo-tetralone to 6-bromo-2-tetralol mediated by the oxidation of NAD(P)H to NAD(P) occurs naturally in the yeast *Trichosporon capitatum*. This reaction was enhanced by electrically regenerating NAD(P)H using the electron shuttle, neutral red. In the presence of a −1.5 V potential the overall reaction rate was increased by 45% [37].

A common research focus for BERs is the detoxification of hazardous chemicals in industrial waste streams and drinking water. One study demonstrated mediated (anthraquinone disulfonate) and nonmediated cathodic perchlorate reduction with cultures of *Dechloromonas* and *Azospira* species. Gas phase analysis suggested that H_2 generated at the cathode might actually function as the electron donor. This approach has an advantage of simplicity over other bioreactor designs that require the continuous addition of a chemical electron donor. The addition of these chemicals requires that their concentration be carefully monitored to prevent excess electron donators from contributing to biofouling and to the formation of trihalomethanes [38].

In another detoxification application, the addition of a redox mediator, methyl viologen, to a mixed anaerobic dechlorinating culture facilitated cathodic reduction of tetrachloroethylene to vinyl chloride. Here, with the electrode poised at −500 mV (vs SHE), addition of methyl viologen resulted in a fivefold increase in current (15–20 μA) and the formation of dechlorination products [39].

3.3 *Living Cell Biosensors*

Living or whole cell biosensors transduce the detection of an analyte to a signal that can be detected electronically. As such, whole-cell biosensors have attracted interest because their relative simplicity, low potential production cost, high sensitivity and selectivity for particular analytes make them competitive with traditional analytical instrumentation. Most biosensors are designed around a specific cellular enzyme that confers selectivity to the target analyte. All of the biosensors described below have cell-free analogues where purified enzymes or other cellular components are coupled to devices as sensing elements. The advantages and disadvantages to the use of whole cell systems as biosensors have been presented in Sect. 2.3. To a great extent the choice of a cell-free or cell/based device is dependent whether or not the environmental conditions of a particular application permit viable cells to be maintained.

3.3.1 Amperometric Biosensors

Amperometric sensors detect current flow between a working and counter electrode induced by a redox reaction at the working electrode. The development of amperometric biosensors began in the 1960s with the immobilization of the enzyme glucose oxidase on a standard oxygen electrode as a glucose sensor.

In a common living cell, amperometric application microorganisms are attached to an amperometric Clark oxygen electrode and the organism's oxygen respiration is measured under various conditions. The most widely used application is for the sensing of biological oxygen demand (BOD) in water contaminated with organic pollutants. The utility of these devices is demonstrated by the fact that more than ten commercial companies sell them [11]. Recent developments in this area have aimed at improving the oxygen transducer [11]. In a recent report a high-throughput BOD sensor was fabricated by applying an organically modified silica oxygen sensing film on polystyrene microtiter plates [40]. The sensing film was formed by embedding the oxygen sensitive dye ruthenium chloride in the silica composite. The bacterium *Stenotrophomonas maltophilia* was immobilized by mixing cultures with the composite prior to coating the plates. BOD values were determined by this device within 20 min, an improvement over the conventional 5-day method [40].

Amperometric microbial biosensors of similar design have been used for the detection of chemicals including ethanol, sugars, phenols and organophosphates.

While these devices demonstrate good sensitivity and stability, they generally demonstrate poor selectivity [11]. An improvement in glucose selectivity was achieved by using ferricyanide as the electron acceptor for *G. oxydans* immobilized on a carbon electrode [41, 42]. A similar approach was used to improve sugar selectivity by using hexacyanoferrate as an electron acceptor for immobilized *G.oxydans* [43]. Attachment of phenol degrading *Pseudomonas putida* to gold electrodes was used to detect phenol [44]. Here *P. putida* was immobilized on the working electrode by using gelatin membrane cross-lined with glutaraldehyde used in conjunction with a Clark oxygen electrode [44].

3.3.2 Potentiometric Biosensors

Potentiometric sensors measure the potential difference between a working electrode and a reference electrode. In this application an ion-selective electrode or a gas-sensing electrode is coated with living microbes. The microbial degradation of an analyte results in ion accumulation that causes a change in potential. In one example an *E. coli* engineered to express organophosphorus hydrolase was attached to a pH electrode by adsorption to a polycarbonate membrane. If organophosphate is present, it is hydrolyzed releasing protons that are detected by the electrode [45]. In another approach a potentiometric oxygen electrode with *Saccharomyces ellipsoideus* immobilized in a dialysis bag was used to determine the concentration of ethanol in beverages. This device yielded a linear calibration curve for ethanol over two orders of magnitude with a response time of 7 min [46]. More recently, *Trichosporon jirovecii* yeast cells were immobilized by membrane retention on a electrode to make a L-cysteine sensor. In the presence of the yeast enzyme L-cysteine desulfhydrase, any L-cysteine in solution is hydrolyzed to pyruvate, ammonia and sulfide ion. The rate of sulfide ion formation is potentiometrically measured over a range of 0.2–150 mg L^{-1} with a detection limit of 1 μM [47].

3.3.3 Impedance-Based Biosensors

The rationale for impedance measurements in cells is based on the fact that, while impedance is directly proportional to resistance and inversely proportional to capacitance, resistance and capacitance are inversely and directly proportional to the free electrode area. When cells (with insulating membranes) grow over an electrode, the free area decreases, resulting in an increase in impedance [48]. Impedance-based devices have, therefore, been used to monitor changes in cell shape. Using a waveform generator, cells on electrodes were subjected to mV amplitude signals in the KHz range and resistance, capacitance and impedance data were collected. The first published application of impedance measurements to the study of cellular function was a study of fibroblasts spreading on an electrode. Cells changed shape from a round to flattened morphology as the cells made greater surface area contact with the electrode and impedance was found to increase [49].

This approach has also been used to monitor changes in cell shape during apoptosis. Apoptosis is characterized by alterations in cell shape due to shrinkage and changes in cell-to-cell contacts. This study monitored apoptosis-induced changes in endothelial cells derived from cerebral microvessels on gold electrodes with a time resolution of minutes using a frequency range of 1–10^6 Hz [50].

Recently, impedance sensing was used to monitor the tightness of cell junctions during the growth of cerebral endothelial cells on gold-film electrodes. Here the electrical resistance of the cell layers was taken as a measure of the integrity of the endothelial barrier. The authors were able to deduce that cerebral cell-derived extracellular matrix material improved the tightness of endothelial cells compared to nonbrain endothelial cells [51].

3.3.4 Cellular Field-Effect Transistors

Integration of biological materials with field effect transistors led to the creation of biologically sensitive field-effect devices (BioFETs). FETs are derived from insulated-gate field effect transistors where the gate electrode is replaced by a test solution and a reference electrode. Inclusion of an ion and/or a charge sensitive gate layer creates a device that is exquisitely sensitive to electrical interactions at the gate [52]. When living cells are applied to the gate, the resulting cellular FETs can be used to detect electrical communication within neurons, transmission paths of ionic channels and as a biosensor detect analytes such as toxic substances (Fig. 5a). FETs have seen wide application in physiological studies several of which are detailed in the next section.

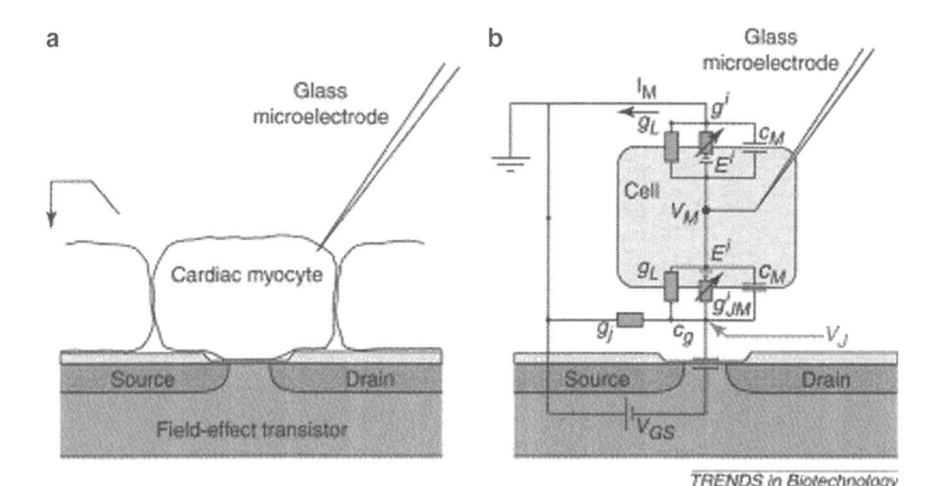

Fig. 5 Electrical equivalent circuit of a cell FET. **a** The cell FET assembly. A monolayer of cardiac myocytes is cultured on the FET surface for both extracellular (FET) and intracellular (microelectrode) recordings. **b** Electrical equivalence circuit for the cell FET. The space between the cell and the gate constituted by the electrolyte has a conductance g_j, the membrane has a capacitance C_M, the gate oxide a capacitance c_g, specific ion conductances g_i, and the extracellular voltage V_M is the voltage measured at the gate. Used by permission from Offenhausser [69]

3.4 *Interfacing of Cells for Physiological Monitoring and Control*

The advantages of interfacing living cells, alluded to in Sect. 2.3, enable noninvasive, real-time monitoring of cellular physiology. By means of different interface designs, the extracellular and intracellular environments of cells can be monitored quantitatively. Approaches may be categorized into two main types, those that are applicable to most types of cells and approaches specific for electrogenic cells such as nerve and heart cells.

3.4.1 General Approaches for all Cell Types

Oxygen consumption and acidification are both considered important parameters to assess cellular response to changes in cellular physiology. The pH environment is an important regulator in cellular physiology and plays an important role in tumor biology and intracellular signaling [53]. During metabolism, acidic metabolic products lower the pH of the intracellular environment which leads to the excretion of protons that can be monitored by ion sensitive microelectrodes and with ion sensitive FETs.

It has been demonstrated that pH is significantly reduced in solid tumors compared with normal cells due to acceleration of glycolytic metabolism [53]. Researchers have measured the pH in real time within 10–100 nm of the cell membranes by using a flow cell within which adherent tumor cells were cultured on an array of pH-ion sensitive FETs. pH measurements of FETs with adherent cells were compared to FETs with no cells bound [53].

In a more recent paper *E. coli* cells were immobilized on agarose gels fixed to an ion-sensitive FET for the purpose of measuring extracellular pH. A quasi-Nernstian pH response was obtained when pH varied from 3 to 12 and the system was stable for several days. Cells were observed to acidify or alkalinize their environment depending on substrate preference [8].

A cellular FET chip has been designed that enables the simultaneous measurement of both pH and O_2 from the same site on a cell. The authors suggest that observation of only one of these two parameters could be misleading and, additionally, that measurement of pH and O_2 on different areas of the cell could also be misleading. They also suggest that their FET design could be further miniaturized such that several sensors could be placed under each individual cell [54].

Another recent approach used *Saccharomyces cerevisiae* immobilized on a carbon dioxide electrode to monitor cellular metabolism. The yeast cells were retained using a silicone gas-permeable membrane. When CO_2 was generated during respiration, it diffused across the membrane and changed the pH of a thin film of electrolyte solution between the electrode and the silicon membrane [55].

A commercial physiological FET-based system has been designed for sophisticated online analysis of the response of living cells to pharmaceuticals. In the Bionas 2500 analyzing system (Bionas GmbH; Rostock, Germany) human cells are grown on a chip that permits acidification rate, oxygen consumption and impedance to be determined in parallel. Buffer flows over the cells at a rate of 56 μL min^{-1} until a reading is taken at which point the flow stops. After adding chemical inhibitors for individual pathways, a decrease in acidification rate was detected. Similarly after adding inhibitors to oxidative phosphorylation, a decrease in respiration was detected. By assessing these parameters the authors suggest that the mode of action of a substance could be determined [56].

3.4.2 Approaches for Electrogenic Cells

Though physiologists have studied the electrical activity of electrogenic cells since the time of Galvani, it was the development of the microelectrode in the 1940s that permitted the electrical study of individual cells. The electrical basis of the nerve impulse was established in 1957 using microelectrodes to study Crayfish neurons [57]. The invention of the patch clamp technique in the 1970s permitted the measurement of ionic currents through single ion channels which established that the movement of ions through gated membrane transporters was the mechanism of the cellular electrical potential. In the patch clamp technique, a micropipette is placed against the cell membrane and a metal electrode inside detects changes in voltage or current.

A further advance in the study of electrogenic cells was the development of microfabricated electrodes in the 1970s using photolithographic methods. Electrode chips were fabricated with apertures <1 μm that replaced the pipette patch clamp. Cells were placed on the aperture in an electrolyte solution. One of the early reports in the 1970s described noninvasive stimulation and recording of the resultant action potentials from heart cells cultured on electrode arrays prepared on glass slides [58]. Later work in the 1980s used microfabricated electrodes to stimulate individual cardiac cells to observe the resultant action potentials [59].

Microelectrodes have been used extensively to study the basic functions that regulate neurotransmitter release both in brain slices and the whole brains of anesthetized animals [60]. Microelectrode studies proved that the exocytotic release of neurotransmitters was the primary means of neurosecretion [61]. In one report carbon fiber microelectrodes were used to detect release of dopamine from dissected substania nigra cells by amperometry. The data indicated that dopamine was released from neuron somata by exocytosis and that this mechanism was regulated by neuronal electrical activity [62].

In all electrode designs using interfaced electrogenic cells, a capacitive or a Faradaic current across the electrode interface causes a voltage gradient in the extracellular space that polarizes the cell leading to an action potential [3]. Quite a bit of work has gone into modeling stimulus parameters and consideration of the potentially damaging effects of electrochemical reaction products alludes to above in Sects. 1 and 2.2 [3].

While the basic approach of establishing neural networks on microfabricated electrodes has continued without much change for over 20 years, the theoretical and conceptual approaches have continued to evolve. The advantage of microfabricated electrodes is that they permit arrays of microelectrodes to be made enabling high-spatiotemporal resolution recording which, in turn, permits the online study of neuronal network plasticity [63]. Recently this approach has been used to study the mechanisms of association formation and learning. Rat hippocampal cells were plated and grown on a commercial 60-electrode array (Multi Channel Systems, Reutlingen, Germany). Hippocampal neuronal networks were subjected to synchronized electrical pulses that caused formation of association between the stimulus and the resultant action potential [64]. Cells were trained using low frequency stimulation (0.2–1.0 Hz). Before learning neurons exhibited random spontaneous electrical bursts; after learning the bursts became synchronized. The authors concluded that firing association and synchrony of spontaneous bursts in the neuronal networks were promoted by learning [64].

While effective, the use of microelectrodes has several disadvantages. First, the positioning of micropipettes is difficult to reproduce from experiment to experiment. Second, the microelectrode has high impedance which results in a poor signal-to-noise ratio [65]. In an attempt to reduce these problems, microelectrode arrays have recently been designed with signal conditioning circuitry associated with each electrode. One such chip, intended to facilitate both stimulation and recording of nerve cells, was composed of 128 electrodes with bandpass filters included in each pixel to limit the noise bandwidth [63].

In another approach to bypass the limitations of microarrays, FETs were designed to monitor the electrical activity of cells occupying the gate region. In one of the first descriptions of the use of *p*-channel FETs for cellular monitoring, the neuron from a leech was attached to the open gate of a FET. It was presumed that the nerve cell was directly coupled to the gate via the oxide. Here stimulated action potentials were found to modulate the source-drain current on the chip [66]. In a later study the authors achieved capacitive stimulation of a leach neuron on a *p*-type FET. In this approach no faradic current flows across the electrode/electrolyte interface in contrast to all systems using a metallic interface thus avoiding the electrochemical reactions described in Sect. 2.1 [67].

Modeling of electrogenic cell behavior on FETs has been described using electrical equivalent circuits [68, 69]. The space between the cell and the gate constituted by the electrolyte has a conductance g_j, the membrane has a capacitance C_M, the gate oxide a capacitance c_g, specific ion conductances g_i, and the extracellular voltage V_M is the voltage measured at the gate [69] (Fig. 5b).

To avoid damaging cells by electroporation or electrolysis, a capacitive stimulating silicon chip was developed for the extracellular stimulation of nerve cells. Using HEK293 cells that overexpressed a Na channel, cells were stimulated by rising and falling voltage ramps without Faradaic current flow. Falling ramps caused a transient sodium inward flow that gave rise to depolarization of the cell [3].

A novel application of ISFETs has been to study of the effects of high-frequency magnetic fields (EMF) on nerve cells. Primary neurons were grown on chips combining a multielectrode array with 58 electrodes and ISFET for pH measurements. The device was placed in an EMF-exposure chamber and cells were exposed to frequencies of 1.56–2.38 GHz. The authors concluded that, because of problems with noise, no statement as to the effects of EMF on neurons could be made [70].

In vitro advances, such as those detailed in this section, have enabled remarkable progress in the field of in vivo brain-machine interfaces (BMI). Projected applications of BMIs include upper and lower limb prostheses for spinal cord injuries and stroke, bladder prostheses, cochlear and brain-stem auditory prostheses, retinal and cortical visual prostheses, cognitive control of assistive devices and deep brain stimulation for Parkinson's disease, epilepsy and depression [22]. One of the first BMIs demonstrated was a direct cortical implant that controlled a robotic manipulator. Rats with electrode implants were trained to move a robotic arm to obtain water [71]. Since then a number of laboratories have demonstrated primate arm reaching and grasping using computer cursors or robotic manipulators [72]. The development of clinically useful devices will require a number of advances, the most important of which is achieving long term (i.e., years) recordings of large populations of neurons without damaging the cells. In addition to dealing with electric current damage, as discussed in Sects. 2.1, 2.2 and 2.7, electrodes will have to be designed to reduce the long term immune response effects outlined in Sect. 2.6. One effective strategy has been to coat electrodes with peptides or proteins that promote nerve cell growth. Some of these approaches include coating electrodes with the cell adhesion proteins collagen and fibronectin [73] and cell adhesion peptides [74]. A related approach to release growth factors to promote nerve cell growth near the electrode was shown to be ineffective [18]. Another approach has been to coat electrodes with conducting polymers of polypyrole effectively to increase the surface of the electrode [75]. Use of immunosuppressant drugs has also been used to reduce the initial immune response and reduce glial scar formation. The use of dexamethasone during the implantation of electrodes into cat brains resulted in marked attenuation of glial scar formation [76]. Recent reports suggest that the use of implanted carbon nanotube fiber electrodes, in addition to providing high surface area, do not initiate an immune response and support attachment, spreading and growth of neurons [77, 78].

4 Concluding Remarks

Considering the accomplishments cited across all disciplines, it is clear that the goal to monitor and control living cells by direct electronic interface, until recently only a subject of science fiction stories, is now being realized. Some of these devices, such as BOD biosensors and the Bionas metabolic analysis system, have found their way out of the laboratory into commercial products. Microbial fuel cells will probably soon find commercial application to power deep sea sensors. As the field of bioelectronics continues to develop we can expect to see technical advances that permit more sophisticated monitoring and control of cellular function. Significant advancement will require (1) more profound understanding of electron transfer mechanisms, (2) design of biocompatible transducer systems that also reduce or eliminate electrolysis or electroporation effects, (3) methods to monitor the condition of electrodes implanted in tissues or attached to cells and (4) better entrapment or implant procedures that enhance the stability of living cells. With a better understanding of these processes it is not too hard to imagine a time when most cellular processes may be monitored or controlled electronically. Realizing the goal of a direct interface with the brain, while in its infancy, is on the horizon. This field has great potential to restore or enable motor function in paralyzed patients. Advancement in these areas will also lead to the development of implantable drug delivery or synthesis devices that will usher in a new paradigm in medical intervention that could eclipse the present pharmacological paradigm.

References

1. Merrill DR, Bikson M, Jefferys JGR (2005) Electrical stimulation of excitable tissue:design of efficacious and safe protocols. J Neurosci Methods 141:171–198
2. Thrash JC, Coates JD (2008) Review: direct and indirect electrical stimulation of microbial metabolism. Environ Sci Technol 42:3921–3931
3. Schoen I, Fromherz P (2008) Extracellular stimulation of mammalian neurons through repetitive activation of Na + channels by weak capacitive currents on a silicon chip. J Neurophysiol 100: 346–357
4. He HQ, Chang DC, Lee YK (2007) Using a micro electroporation chip to determine the optimal physical parameters in the uptake of biomolecules in HeLa cells. Bioelectrochemistry 70: 363–368
5. Ryttsen F, Farre C, Brennan C, Weber SG, Nolkrantz K, Jardemark K, Chiu DT, Orwar O (2000) Characterization of single-cell electroporation by using patch-clamp and fluorescence microscopy. Biophys J 79:1993–2001
6. Olofsson J, Nolkrantz K, Ryttsen F, Lambie BA, Weber SG, Orwar O (2003) Single-cell electroporation. Curr Opin Biotechnol 14:29–34
7. Debabov VG (2008) Electricity from microorganisms. Microbiology 77:123–131
8. Bettaieb F, Ponsonnet L, Lejeune P, Ben Ouada H, Martelet C, Bakhrouf A, Jaffrezic-Renault N, Othmane A (2007) Immobilization of E-coli bacteria in three-dimensional matrices for ISFET biosensor design. Bioelectrochemistry 71:118–125
9. D'Souza SF (2001) Microbial biosensors. Biosens Bioelectron 16:337–353
10. Wen G, Zheng J, Zhao C, Shuang S, Dong C, Choi MMF (2008) A microbial biosensing system for monitoring methane. Enzyme Microb Technol 43:257–261
11. Lei Y, Chen W, Mulchandani A (2006) Microbial biosensors. Anal Chim Acta 568:200–210
12. Lojou, E, Bianco P (2006) Application of the electrochemical concepts and techniques to amperometric biosensor devices. J Electroceram 16:79–91
13. Premkumar JR, Lev O, Marks RS, Polyak B, Rosen R, Belkin S (2001) Antibody-based immobilization of bioluminescent bacterial sensor cells. Talanta 55:1029–1038
14. Kirgoz UA, Timur S, Odaci D, Perez B, Alegret S, Merkoci A (2007) Carbon nanotube composite as novel platform for microbial biosensor. Electroanalysis 19:893–898
15. Wilson CJ, Clegg RE, Leavesley DI, Pearcy MJ (2005) Mediation of biomaterial-cell interactions by adsorbed proteins: a review. Tissue Eng 11:1–18
16. Kim L, Toh YC, Voldman J, Yu H (2007) A practical guide to microfluidic perfusion culture of adherent mammalian cells. Lab Chip 7:681–694
17. Nicolelis MAL, Dimitrov D, Carmena JM, Crist R, Lehew G, Kralik JD, Wise SP (2003) Chronic, multisite, multielectrode recordings in macaque monkeys. Proc Natl Acad Sci U S A 100:11041–11046

18. Polikov VS, Tresco PA, Reichert WM (2005) Response of brain tissue to chronically implanted neural electrodes. J Neurosci Methods 148:1–18
19. Stroscio MA, Dutta M (2005) Integrated biological-semiconductor devices. Proc IEEE 93: 1772–1783
20. Purves D, Augustine GJ, Fitzpatrick D, Hall WC, LaMantia AS, McNamara JO, White LE (2004). Neuroscience. Sinauer Associates, Sunderland, MA
21. Brummer SB, Turner MJ (1977) Electrochemical considerations for safe electrical-stimulation of nervous-system with platinum-electrodes. IEEE Trans Biomed Eng 24:59–63
22. Cogan SR (2008) Neural stimulation and recording electrodes. Annu Rev Biomed Eng 10: 275–309
23. Potter MC (1911) Electrical effects accompanying the decomposition of organic compounds. Proc R Soc Lond B Biol Sci 84:260–276
24. Rabaey K, Rodriguez J, Blackall LL, Keller J, Gross P, Batstone D, Verstraete W, Nealson KH (2007) Microbial ecology meets electrochemistry: electricity-driven and driving communities. ISME J 1:9–18
25. Clauwaert P, Aelterman P, Pham TH, De Schamphelaire L, Carballa M, Rabaey K, Verstraete W (2008) Minimizing losses in bio-electrochemical systems: the road to applications. Appl Microbiol Biotechnol 79:901–913
26. Tender LM, Gray SA, Groveman E, Lowy DA, Kauffman P, Melhado J, Tyce RC, Flynn D et al. (2008) The first demonstration of a microbial fuel cell as a viable power supply: powering a meteorological buoy. J Power Sources 179:571–575
27. Rabaey K, Lissens G, Siciliano SD, Verstraete W (2003) A microbial fuel cell capable of converting glucose to electricity at high rate and efficiency. Biotechnol Lett 25: 1531–1535
28. Ringeisen BR, Henderson E, Wu PK, Pietron J, Ray R, Little B, Biffinger JC, Jones-Meehan JM (2006) High power density from a miniature microbial fuel cell using Shewanella oneidensis DSP10. Environ Sci Technol 40:2629–2634
29. Lovley, DR (2006) Bug Juice: harvesting electricity with microorganisms. Nature Rev. Micro. 4:497–508.
30. Lies DP, Hernandez ME, Kappler A, Mielke RE, Gralnick JA, Newman DK (2005) Shewanella oneidensis MR-1 uses overlapping pathways for iron reduction at a distance and by direct contact under conditions relevant for biofilms. Appl Environ Microbiol 71:4414–4426
31. Busalmen JP, Esteve-Nunez A, Berna A, Feliu JM (2008) C-type cytochromes wire electricity-producing bacteria to electrodes. Angew Chem Int Edit 47:4874–4877
32. Nevin KP, Lovley DR (2000) Lack of production of electron-shuttling compounds or solubilization of Fe(III) during reduction of insoluble Fe(III) oxide by Geobacter metallireducens. Appl Environ Microbiol 66:2248–2251
33. Lovley DR (2008) Extracellular electron transfer: wires, capacitors, iron lungs, and more. Geobiology 6:225–231
34. Gregory KB, Bond DR, Lovley DR (2004) Graphite electrodes as electron donors for anaerobic respiration. Environ Microbiol 6:596–604
35. Sadoff HL, Halvorson HO, Finn RK (1956) Electrolysis as a means of aerating submerged cultures of microorganisms. Appl Microbiol 4:164–170
36. Park DH, Zeikus JG (1999) Utilization of electrically reduced neutral red by Actinobacillus succinogenes: physiological function of neutral red in membrane-driven fumarate reduction and energy conservation. J Bacteriol 181:2403–2410
37. Shin HS, Jain MK, Chartrain M, Zeikus JG (2001) Evaluation of an electrochemical bioreactor system in the biotransformation of 6-bromo-2-tetralone to 6-bromo-2-tetralol. Appl Microbiol Biotechnol 57:506–510
38. Thrash JC, Van Trump JI, Weber KA, Miller E, Achenbach LA, Coates JD (2007) Electrochemical stimulation of microbial perchlorate reduction. Environ Sci Technol 41:1740–1746
39. Aulenta F, Catervi A, Majone M, Panero S, Reale P, Rossetti S (2007) Electron transfer from a solid-state electrode assisted by methyl viologen sustains efficient microbial reductive dechlorination of TCE. Environ Sci Technol 41:2554–2559

40. Pang HL, Kwok HY, Chan PH, Yeung CH, Lo WH, Wong KY (2007) High-throughput determination of biochemical oxygen demand (BOD) by a microplate-based biosensor. Environ Sci Technol 41:4038–4044
41. Tkac J, Vostiar I, Gemeiner P, Sturdik E (2002) Monitoring of ethanol during fermentation using a microbial biosensor with enhanced selectivity. Bioelectrochemistry 56:127–129
42. Tkac J, Vostiar I, Gorton L, Gemeiner P, Sturdik E (2003) Improved selectivity of microbial biosensor using membrane coating. Application to the analysis of ethanol during fermentation. Biosens Bioelectron 18:1125–1134
43. Held M, Schuhmann W, Jahreis K, Schmidt HL (2002) Microbial biosensor array with transport mutants of Escherichia coli K12 for the simultaneous determination of mono-and disaccharides. Biosens Bioelectron 17:1089–1094
44. Timur S, Pazarlioglu N, Pilloton R, Telefoncu A (2003) Detection of phenolic compounds by thick film sensors based on Pseudomonas putida. Talanta 61:87–93
45. Mulchandani, A, Mulchandani P, Chauhan S, Kaneva I, Chen W (1998) A potentiometric microbial biosensor for direct determination of organaphosphate nerve agents. Electroanalysis 10:733–737
46. Rotariu L, Bala C, Magearu V (2004) New potentiometric microbial biosensor for ethanol determination in alcoholic beverages. Anal Chim Acta 513:119–123
47. Hassan SSM, El-Baz AE, Abd-Rabboh HSM (2007) A novel potentiometric biosensor for selective L-cysteine determination using L-cysteiine-desulfhydrase producing Trichosporon jirovecii yeast cells coupled with sulfide electrode. Anal Chim Acta 602:108–113
48. Spegel C, Heiskanen A, Skjolding LHD, Emneus J (2008) Chip based electroanalytical systems for cell analysis. Electroanalysis 20:680–702
49. Giaever I, Keese CR (1991) Micromotion of mammalian-cells measured electrically. Proc Natl Acad Sci U S A 88:7896–7900
50. Arndt S, Seebach J, Psathaki K, Galla HJ, Wegener J (2004) Bioelectrical impedance assay to monitor changes in cell shape during apoptosis. Biosens Bioelectron 19:583–594
51. Hartmann C, Zozulya A, Wegener J, Galla HJ (2007) The impact of glia-derived extracellular matrices on the barrier function of cerebral endothelial cells: an in vitro study. Exp Cell Res 313:1318–1325
52. Schoning MJ, Poghossian A (2006) Bio FEDs (field-effect devices): state-of-the-art and new directions. Electroanalysis 18:1893–1900
53. Lehmann M, Baumann W, Brischwein M, Ehret R, Kraus M, Schwinde A, Bitzenhofer M, Freund I, Wolf B (2000) Non-invasive measurement of cell membrane associated proton gradients by ion-sensitive field effect transistor arrays for microphysiological and bioelectronical applications. Biosens Bioelectron 15:117–124
54. Lehmann M, Baumann W, Brischwein M, Gahle HJ, Freund I, Ehret R, Drechsler S, Palzer H et al (2001) Simultaneous measurement of cellular respiration and acidification with a single CMOS ISFET. Biosens Bioelectron 16:195–203
55. Martinez M, Hilding-Ohlsson A, Viale AA, Corton E (2007) Membrane entrapped Saccharomyces cerevisiae in a biosensor-like device as a generic rapid method to study cellular metabolism. J Biochem Biophys Methods 70:455–464
56. Thedinga E, Kob A, Holst H, Keuer A, Drechsler S, Niendorf R, Baumann W, Freund I et al. (2007) Online monitoring of cell metabolism for studying pharmacodynamic effects. Toxicol Appl Pharmacol 220:33–44
57. Teulon J (2004) Patch-clamp techniques. M S-Med Sci 20:550
58. Thomas CA, Springer PA, Okun LM, Berwaldn Y, Loeb GE (1972) Miniature microelectrode array to monitor bioelectric activity of cultured cells. Exp Cell Res 74:61–69
59. Israel D, Barry WH, Edell DJ, Mark RG (1984) An array of microelectrodes to stimulate and record from cardiac cells in culture. Am J Physiol 247:H669–H674
60. Dale N, Hatz S, Tian FM, Llaudet E (2005) Listening to the brain: microelectrode biosensors for neurochemicals. Trends Biotechnol 23:420–428
61. Pothos EN, Davila V, Sulzer D (1998) Presynaptic recording of quanta from midbrain dopamine neurons and modulation of the quantal size. J Neurosci 18:4106–4118

62. Jaffe EH, Marty A, Schulte A, Chow RH (1998) Extrasynaptic vesicular transmitter release from the somata of substantia nigra neurons in rat midbrain slices. J Neurosci 18:3548–3553
63. Heer F, Hafizovic S, Franks W, Blau A, Ziegler C, Hierlemann A (2006) CMOS microelectrode array for bidirectional interaction with neuronal networks. IEEE J Solid State Circuits 41: 1620–1629
64. Li YL, Zhou W, Li XN, Zeng SQ, Liu M, Luo QM (2007) Characterization of synchronized bursts in cultured hippocampal neuronal networks with learning training on microelectrode arrays. Biosens Bioelectron 22:2976–2982
65. Kovacs GTA (2003) Electronic sensors with living cellular components. Proc IEEE 91:915–929
66. Fromherz P, Offenhausser A, Vetter T, Weis J (1991) A neuron-silicon junction – a retzius cell of the leech on an insulated-gate field-effect transistor. Science 252:1290–1293
67. Fromherz P, Stett A (1995) Silicon-neuron junction – capacitive stimulation of an individual neuron on a silicon chip. Phys Rev Lett 75:1670–1673
68. Vassanelli S, Fromherz P (1998) Transistor records of excitable neurons from rat brain. Appl Phys A Mater Sci Process 66:459–463
69. Offenhausser A, Knoll W (2001) Cell-transistor hybrid systems and their potential applications. Trends Biotechnol 19:62–66
70. Koester P, Sakowski J, Baumann W, Glock HW, Gimsa J (2007) A new exposure system for the in vitro detection of GHz field effects on neuronal networks. Bioelectrochemistry 70:104–114
71. Chapin JK, Moxon KA, Markowitz RS, Nicolelis MAL (1999) Real-time control of a robot arm using simultaneously recorded neurons in the motor cortex. Nat Neurosci 2:664–670
72. Lebedev MA, Nicolelis MAL (2006) Brain-machine interfaces: past, present and future. Trends Neurosci 29:536–546
73. Ignatius MJ, Sawhney N, Gupta A, Thibadeau BM, Monteiro OR, Brown IG (1998) Bioactive surface coatings for nanoscale instruments: effects on CNS neurons. J Biomed Mater Res 40:264–274
74. Kam L, Shain W, Turner JN, Bizios R (2002) Selective adhesion of astrocytes to surfaces modified with immobilized peptides. Biomaterials 23:511–515
75. Yang JY, Martin DC (2004) Microporous conducting polymers on neural microelectrode arrays – I – Electrochemical deposition. Sens Actuator B Chem 101:133–142
76. Maynard EM, Fernandez E, Normann RA (2000) A technique to prevent dural adhesions to chronically implanted microelectrode arrays. J Neurosci Methods 97:93–101
77. Gheith MK, Sinani VA, Wicksted JP, Matts RL, Kotov NA (2005) Single-walled carbon nanotube polyelectrolyte multilayers and freestanding films as a biocompatible platformfor neuroprosthetic implants. Adv Mater 17:2663
78. Dubin RA, Callegari GC, Kohn J, Neimark AV (2008) Carbon nanotube fibers are compatible with mammalian cells and neurons. IEEE Trans Nanobiosci 7:11–14

Adv Biochem Engin/Biotechnol (2010) 117: 179–191
DOI: 10.1007/10_2009_1

Published online: 19 June 2009

On-Chip Detection of Cellular Activity

R. Almog, R. Daniel, S. Vernick, A. Ron, H. Ben-Yoav, and Y. Shacham-Diamand

Abstract The use of on-chip cellular activity monitoring for biological/chemical sensing is promising for environmental, medical and pharmaceutical applications. The miniaturization revolution in microelectronics is harnessed to provide on-chip detection of cellular activity, opening new horizons for miniature, fast, low cost and portable screening and monitoring devices. In this chapter we survey different on-chip cellular activity detection technologies based on electrochemical, bio-impedance and optical detection. Both prokaryotic and eukaryotic cell-on-chip technologies are mentioned and reviewed.

Keywords Whole-cell • Biochips • Biosensors • Electrochemical detection • Bioluminescence

Contents

R. Almog (✉), R. Daniel, S. Vernick, A. Ron, H. Ben-Yoav, Y. Shacham-Diamand
Department of Physical Electronics, School of Electrical Engineering,
Faculty of Engineering, Tel-Aviv University, Tel-Aviv 69978, Israel
e-mail: almogr@tx.technion.ac.il

1 Introduction

The miniaturization of biosensors is a fast growing field of research, due to the microelectronics revolution together with advancement in microfluidics and the possibility to immobilize live cells on a chip. Cell based screening is a powerful method that uses living cells to test the effect of different molecules like drugs and toxicants on the cellular and molecular phenotype of cells. This technique requires the use of highly sensitive tools that permit high speed systematic identification of biochemical targets and markers on given cell libraries. Electrical sensors can be miniaturized down to a cell-size scale and be positioned directly at the vicinity of the cell surface, where cellular signaling substances are captured before they diffuse.

The use of on-chip cellular activity monitoring as a biological sensing mechanism is promising for environmental, medical and pharmaceutical applications, yielding smaller, faster and cheaper biosensors [1]. These biosensors are generally referred as "whole-cell" biosensors [2–4] and they usually use microbes (prokaryotic cells) as functional sensors for environmental applications or eukaryotic cells for health care and medical applications.

Microbes can generate a variety of specific responses to chemical and biological stimulations. In the biotic-micro-electro-mechanical systems (biotic-MEMS) sensor, integrating live microbial cells with microfabricated structures, the main sensing element is the microbe that converts the chemical or biological signal to typically an electrical one [5]. Microbial sensors are less sensitive than enzyme-based sensors to environmental changes; however, the microbes can be genetically engineered to have an altered response [6]. Microbial biosensors have been utilized for environmental monitoring [7], food monitoring, e.g., sensing of alcohol [8], glucose [9] or fatty acids [10], monitoring of microbial growth rate [11] and biocide measurements [12].

There are two general approaches to the monitoring of chemicals in the environment. The traditional approach is based on chemical or physical analysis and allows highly accurate and sensitive determination of the exact composition of the sample. This approach is highly specific, complex, and requires specialized laboratories. On the other hand, using live cells as the biological sensing entity is nonspecific and can be used to detect, by very simple means, complex series of reactions that can exist only in an intact, functioning cell [13].

Another class of whole-cell biosensors is based on electrically active living cells cultured on extracellular electrode arrays and can be utilized to detect biologically active agents [14]. Neurons can be immobilized on microfabricated surfaces when changes in their electrical signals upon exposure to harmful chemicals have been measured on a chip [15].

In this chapter we describe whole-cell sensing methods based on electrochemical detection, electrical impedance spectroscopy, electrochemical enzyme-linked-immunosorbant assay (ELISA) and finally on-chip systems with optical sensing.

2 On-Chip Electrochemical Detection of Cellular Activity

Electrochemical biosensors are based on the fact that, during a bio-interaction process, electrochemical species are consumed or generated producing an electrochemical signal which may be measured by an electrochemical detector. Electrochemical measurements only detect the electrical properties of the analyte species undergoing redox reactions, so they are limited to electroactive species. There are three main types of methods for electrochemical detection: amperometry, potentiometry and conductometry (impedance measurement) [16].

Many different electrochemical methods have been studied and applied to cellular bio-sensing. Dissolved oxygen (DO) electrodes are a famous example of cellular bio-sensing. The DO electrodes can be employed to measure the biological oxygen demand (BOD) from an entrapped microfungus [17], or for drug resistance and HACCP (hazard analysis and critical control point) tests in the food industry [18]. Scanning electrochemical microscopy (SECM) is a new technology derived from scanning probe microscopy (SPM) using microfabricated electrodes as scanning probes [19]. Using this technology, localization of respiratory activity can be visualized within a single cell surface. Such a respiration pattern may provide new information for cellular signaling and physiology. Cell proliferation can also be utilized to evaluate cell behavior. In situ monitoring of cell proliferation has been reported with a specialized arrayed ISFET [20].

Specific cellular bio-sensing can be utilized by monitoring specific cellular signals unique to a desired analytical situation in a cell. Such key signal substances which are specific to cell behaviors include intracellular nitric oxide (NO), which plays a key role as a molecular messenger in biological systems [21, 22], and cellular protein expression.

Popovtzer et al. [23] presented a nano-biochip, which contains an array of nano volume electrochemical cells, based on silicon micro-system-technology (MST). Genetically engineered *E. coli* bacteria are integrated into the chip to express electrochemically detectable signals in the presence of toxicants.

3 Bio-Impedance

Bio-impedance spectroscopy had been a powerful tool to investigate the passive electrical properties of organelles, cells and tissue for almost 100 years. The electrical properties of biological samples were explored for the first time by Hoeber who investigated the electrical properties of *Erythrocyte* cells [24].

The foundations of bio-impedance technique have been developed by Schwan according to the theories founded earlier by Maxwell and Wagner [25, 26]. Since then, the electrical properties of biological samples have been extensively studied by a number of researchers.

Bioelectrical impedance analysis measures the impedance as a function of the current frequency. In practice, a small constant alternating current is passed between electrodes spanning the biological sample and the voltage between electrodes provides a measure of impedance [34]. This is done over a wide range of frequencies. When the frequency response varies significantly between different biomaterials, detection of physiological or pathological processes is possible.

This method [27] has been used to detect cell physiology changes, e.g., metabolic activity [28], cellular growth and cytotoxicity [29-33]. The method is attractive due to its simplicity and usability since a specialized reference electrode is not needed.

The electrical properties of biological tissues can provide rare insight about their physiological and pathological state. Most tissues display extremely high admittance at low frequencies, falling off in more or less distinct steps with increasing frequency. Their frequency dependence permits identification and investigation of a number of several underlying mechanisms, which are found to be well adapted to the physiological and structural state of the analyzed sample.

The electrical properties of biological tissues are characterized by three major dispersion regions, termed α, β and γ (Fig. 1), each due to a different relaxation mechanism [34], accounting for low frequency, radio frequency, and microwave frequency, respectively. Tissues and cells will typically exhibit a significant dispersion in the radio frequency range. The dispersion caused by proteins and lipids has a much smaller magnitude than cells and tissues, and the characteristic frequency is typically higher [35].

The impedance of biological tissue comprises two components, resistance and reactance. The resistive component is related to the fluids within the tissue which accounted for both extra and intracellular components. The reactance component is attributed to the phospholipids membranes which act as imperfect capacitors, and contribute a frequency dependent reactive component. At low frequency, the conductive charging of membranes is producing a large polarization across the cells and does not allow the current to pass through them. When the frequency is increased such that it exceeds the inverse RC time of the medium–membrane interface,

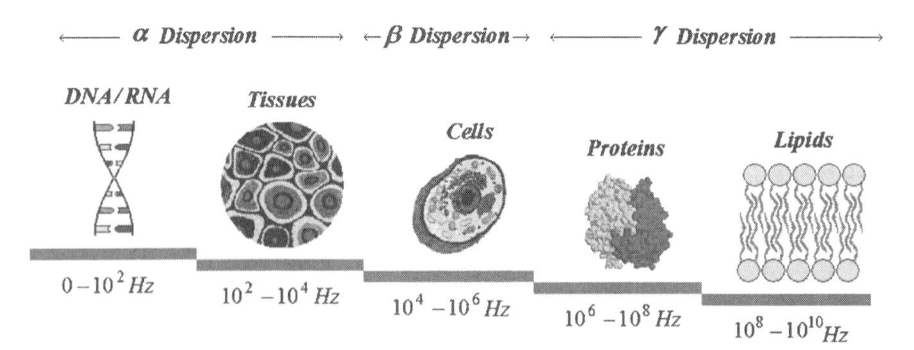

Fig. 1 Frequency dependence of biology samples. The dependence is characterized by α, β and γ dispersions, each due to a different mechanism

current penetration into the cells interior will occur, resulting in dispersion (relaxation) that is governed by cytoplasm (cell interior) and medium conductivities. The ability to separate the extra- and intracompartments allows distinguishing between the two phases related to the molecular and cellular properties of the tissue. This approach was found to be very useful in several applications including calculation of mass, volume and impedance tomography techniques [36-38].

In order to correlate between the measured impedance characteristics and the physiological phenomena occurring in cells or tissues, a matched model should be applied. The electrical and morphological parameters of the tested sample can be determined by fitting the calculated impedance spectrum to the measured spectrum of the tissue. The extracted parameters can be correlated to the given physiological state of the analyzed sample. When changes occur within the sample, those parameters are expected to be adapted to the new physiological state and hence can indicate on several pathological phenomena [34].

Several factors such as the structural geometry, volume and conductivity all have an effect on the measured/calculated impedance and therefore should be taken into account.

The impedance is based on the solution of electrostatic problem within a conducting and dielectric medium (lossy medium) and can be derived using the continuity equation for charge which represents the complex potential of the medium:

$$\nabla(\sigma\nabla u)+\frac{\partial}{\partial t}\left(\nabla(\varepsilon\nabla u)\right)=0, \tag{1}$$

where u is the electric potential, ε is the permittivity, and σ represents the conductivity of the medium. After setting the potential, the impedance spectrum of the sample can be calculated based on the average current density:

$$\frac{1}{Z}=\frac{J}{V}=\frac{J_{\mathrm{C}}+J_{\mathrm{D}}}{V}=\frac{k^{*}E}{V}, \tag{2}$$

where V is the applied potential, J the current density, J_{C} the conduction current, J_{D} the displacement current, E the applied electric field between the electrodes and k the complex conduction factor. The given expression allows calculation of the exact impedance spectrum of any given tissue. It takes into account both structural and electrical properties and hence permits identification of fine and specific changes which are related to the physiological state of the analyzed sample.

3.1 Screening Analysis Using Bio-Impedance Spectroscopy

Bio-impedance spectroscopy was already demonstrated to be very efficient tool for bio-markers screening analysis [39]. In this study, *Madin-Darby Canine Kidney Epithelial* (*MDCK*) and *Bone Marrow derived Preosteoblastic* (*MBA*) cell lines have been used to investigate the effectiveness of impedance spectroscopy after intracellular and membranal markers respectively. Here the intracellular effect was checked using *SEC13* over-expressing *MDCK* cells. Effect of membrane was tested by coating *MBA* cells with free *lectin* antibodies that bind glycosylate elements on cell surface.

Comparing the complex spectra of *MBA WT* (*Wild Type*, untreated cells) cells and the *lectin* coated cells (Fig. 2a) reveals clear differences between those two cell lines. The dielectric constant (ε') obtained for the treated cells is found to be higher than the *WT* cells. This rise is expected due to the presence of charged groups on the membrane, which affect its capacitive part. Additionally, based on the loss (ε'') spectrum (Fig. 2d), changes are also obtained on the frequency dispersion characteristic. Here the loss factor is shifted to lower frequencies due to the increase of membrane

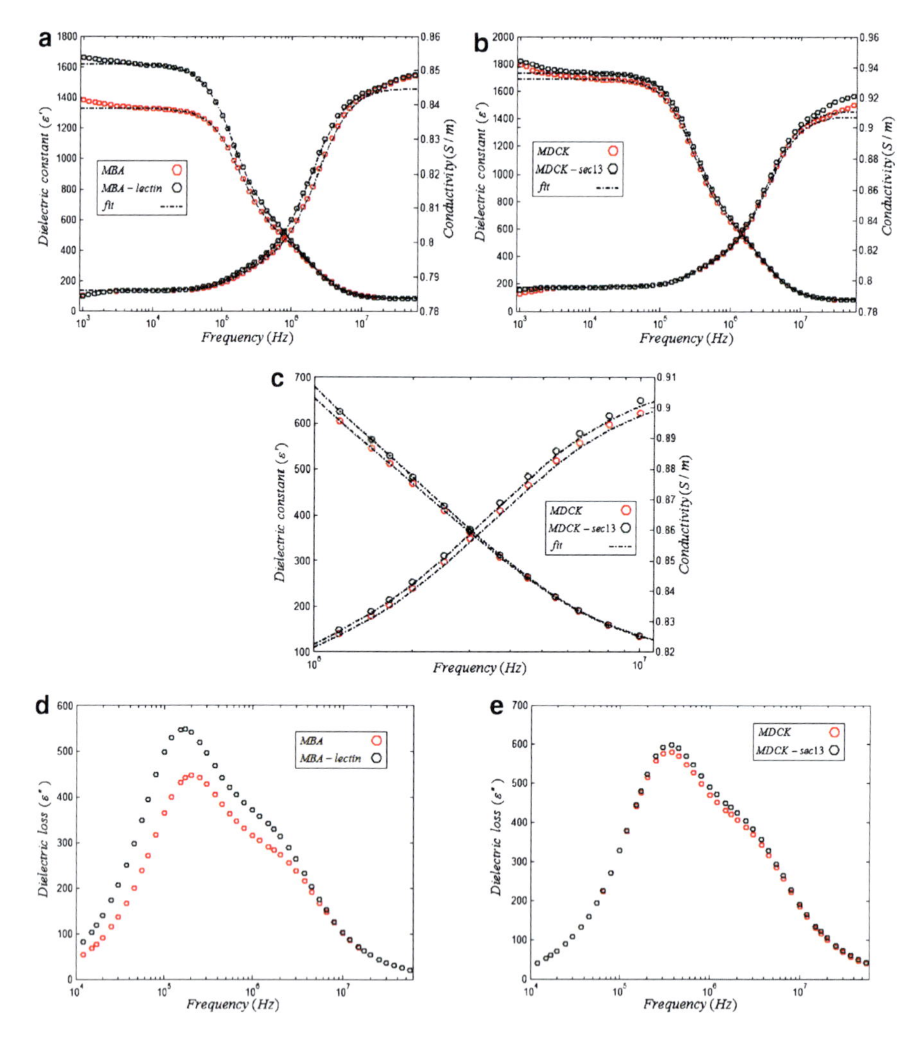

Fig. 2 Dielectric spectra of *MBA* and *MDCK* cell suspensions. **a** Dielectric and conductivity spectra of *MBA* and *MBA-lectin* suspensions. **b** Dielectric and conductivity spectra of *MDCK* and *MDCK-sec13* suspensions. **c** High frequency observation on *MDCK* and *MDCK-sec13* spectra. **d** Loss spectra of *MBA* and *MBA-lectin* suspensions. **e** Loss spectra of *MDCK* and *MDCK-sec13* suspensions

capacitance. This is caused due to change on polarization relaxation time of cells: $f_{ment\text{–}cvt} = 1/\tau_{ment\text{–}cvt}$ which is affected by both membrane capacitance and cytoplasm conductivity ($\tau_{mem\text{–}cyt}$ increased).

The *MDCK* spectra obtained (Fig. 2b, c) demonstrate remarkable differences at relatively high frequencies when charge and ion migration (conductive polarization) start to fade. This happens when the frequency exceeds the inverse relaxation time ($\tau_{med\text{–}mem}$) of the medium–membrane interface. Here the *SEC13* cells exhibit higher dispersion characteristics compared to the *WT* cells. This is caused by changes occurred in cytoplasm conductivity which affects the cytoplasm–membrane relaxation time. The characteristic frequency of the cells ($\tau_{mem\text{–}cyt}$) now increases due to the decrease in cytoplasm resistivity.

The findings presented emphasize the high correlation that exists between molecular and cellular markers and the matching impedance spectra and electrical properties of biological cells. The high sensitivity of impedance spectroscopy as demonstrated here as well as by others in the past has promise potential as a screening tool on live cell libraries. It can be used for quantitative and systematic analysis on multiple microlibraries labeled with specific probes in non invasive and non destructive way. It can also allow real time screening after fine biochemical effects on both the cellular and molecular level, which cannot be detected using traditional screening assays.

4 An Integrated Multimethod Biosensor

Another approach to whole-cell biosensors is to integrate few methods on the same chip. Since whole-cell bio sensing is in principle functional sensing, multiple methods will increase the sensitivity and reduce the false positive and false negative probabilities. For example, we present here a multiple method approach, which integrates electrochemical sensing and modified ELISA for the purpose of early cancer detection. In the field of cancer diagnostics, early detection has a dramatic effect on mortality. The sooner cancer is diagnosed the higher the survival probability of the patient. However, for many types of cancer the current diagnostic methods are insufficient for that purpose.

It is well known that cancer is essentially the result of various cellular pathways and the sum of many events manifesting in different ways. Moreover, some cancers are even harder to diagnose since they express markers which are not necessarily unique (e.g., colon cancer). Therefore, it would be advisable to try to probe the tumor with few methods and gather information which represents various pathways. Accordingly, multitarget strategies have been offered as a suitable solution holding an enormous potential in increasing sensitivity, maintaining specificity and allow for a more accurate, highly confident early diagnostics [40].

In this section, a multimethod integrated biosensor is presented, implementing the above-mentioned concept and providing a way to sense at least four different mechanisms which characterize cancer cells, thus enhancing the confidence level exceedingly.

The multimethod electrochemical biosensor is capable of performing two different diagnostic tests on the same sample. The sample could be a small biopsy or a cell culture derived from it. The first test is a substrate-aided electrochemical detection of enzyme secretion. This test involves the electrochemical distinction between cancerous and normal cells relying on the observed downregulation or upregulation of certain intracellular enzymes by the cancerous cells [41]. These enzymes may also work on substrates which subsequently become electroactive. This approach offers rapid and straightforward detection. However, relying on this detection method solely might lead to undesired false negative/positive results. The relatively high rate of false negative/positive results is a well known issue in cancer diagnostics and decreasing this rate of false results has been a longstanding goal. Therefore, a second diagnostic test is performed on the same sample offering higher specificity. This method could be described as electrochemical immunodetection of specific cancer secreted biomarkers [42]. It is essentially a modified ELISA performed on a suitable chemically modified transducing element.

The inherent specificity of immunoreaction, along with high sensitivity and convenience of various physical transducers, render immunosensors as a major development in the immunochemical field and clinical diagnosis. The amperometric immunosensors were constructed based on the electrochemically active species produced by enzyme labels. Electrochemical detection of the labels has several advantages such as high sensitivity and low cost of the resulting sensors and instrumentation.

Clinical analysis requires methods of high reliability and immunoreactions are recognized for their high sensitivity and selectivity. Antibodies are considered to be well-suited recognition elements for bioassays and immunosensors. Advances in molecular biology have led to much understanding of potential biomarkers that can be used for diagnosis. Antibody-based technology takes advantage of specific interactions with antigenic regions of an analyte to achieve high confidence level and selectivity. The high specificity and affinity of an antibody for its antigen allows a selective binding of the analyte (antigen) which is present in the nano- to picomolar range in the presence of hundreds of other substances, even if they exceed the analyte concentration by two to three orders of magnitude. Amperometric transducers which are used much more frequently than other transducers in electrochemical indirect immunoassays allow the fast detection of currents over a broad linear range with detection limits as low as 10^{-10} A by using commercial potentiostats. Since antibodies and antigens are usually not electrochemically active within the desired potential range, redox-active compounds can be applied as labels for indication.

While the first method represents a typical cancerous cellular mechanism the second method covers different aspects. In this method the detection of up to three biomarkers which represents three different mechanisms on the same platform is feasible. Combination of the methods along with integration of the obtained data may significantly enhance the level of confidence in cancer diagnosis by encompassing various cancerous mechanisms and providing added value that could not otherwise be achieved.

The first detection method is performed on an electrochemical cell including working, reference and auxilary screen-printed electrodes.

Amperometric measurement is employed in order to detect the expressed enzyme. The detection of the desired enzyme is possible due to its biocatalytic activity on the added substrate. The substrate is injected into the electrochemical cell (housed in a chamber) and is capable of diffusing into the cancer cells. The product of the enzymatic reaction is oxidized on the electrode at a typical voltage and the resulting current is measured.

An example is the detection of the enzyme alkaline phosphatase. Few subtypes of this enzyme exist in the human body, and some are disregulated during certain pathologies and malignancies. The inherent biocatalytic activity of this enzyme is dephosphorylation and it is highly efficient in dephosphorylating the substrate *p*-aminophenyl phosphate (pAPP). The result of this process is that the detection of intracellular alkaline phosphatase may be achieved simply by adding pAPP to certain cancer cell solution or biopsy – the substrate diffuses inside the cell and the product, *p*-aminophenol, diffuses outside the cancer cell and is oxidized on a working electrode when a voltage of 0.22 V is applied. Figure 3 illustrates the first method.

The second detection method is also performed on an electrochemical cell including working, reference and auxiliary screen-printed electrodes. The detection of secreted biomarkers is carried out essentially by adjusting the "sandwich-type"

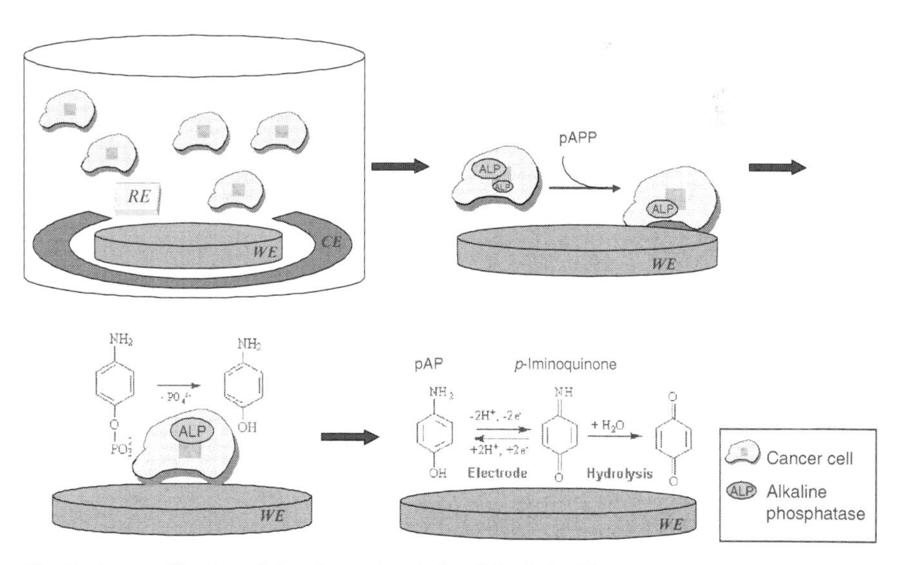

Fig. 3 A magnification of the electrochemical cell loaded with cancer cells (*WE*, *RE* and *CE* refers to working, reference and counter electrode, respectively). The electrochemical substrate, *p*-aminophenyl phosphate (pAPP) is added and subsequently dephosphorilated by the enzyme ALP (alkaline phosphatase). The resulting product, *p*-aminophenol (pAP), is oxidized on the electrode thus generating a measurable current [43]

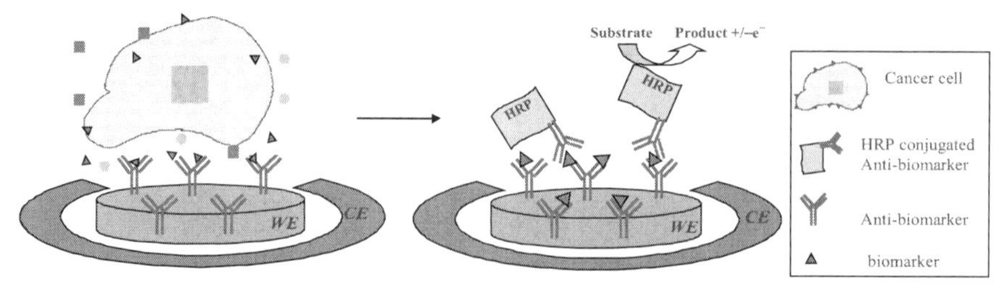

Fig. 4 A magnification of the electrochemical cell loaded with cancer cells (*WE, RE* and *CE* refers to working, reference and counter electrode, respectively). A specific biomarker, typical to cancer cells is secreted and captured by an antibody immobilized to the working electrode. Following removal of cells an immunoconjugate is introduced and binds the captured biomarker. Following removal of unbound immunoconjugate electrochemical substrate is introduced and catalyzed by the enzyme label resulting in a measurable current [43]

ELISA method – a traditionally employed bio-analytical method. Antibodies against three different biomarkers are immobilized on three different electrodes to form an array. Cell medium containing the secreted biomarkers (or the cell culture itself) is incubated (or flowed through) in the electrochemical chamber. Following the capture of the biomarkers by the immobilized antibodies, an immunoconjugate solution is introduced, which is essentially the same antibody labeled with an enzyme capable of eliciting an electrochemical reaction with a substrate (e.g., horseradish peroxidase or alkaline phosphatase). After washing the unbound immunoconjugate the substrate is added and the resulting current is measured. Figure 4 illustrates the second method.

The electrochemical ELISA-like method is highly sensitive; while each antigen could not be considered as a reliable marker by itself, the detection of three increases the confidence dramatically.

Altogether this multimethod biosensor deals with the simultaneous detection of cancer from inside the cell (intracellular enzyme expression) to the extracellular part (secreted markers).

5 Monitoring On-Chip Cellular Activity by Optical Detection

Environmental biosensors make use of genetically engineered *E. coli* bioluminescent biosensors which emit light when exposed to toxic materials [44, 45]. The illumination intensity and its time dependence is a function of the toxicants' dose and type. As an example of this method (see [46]) *E. coli* cells were genetically engineered, carrying a recA::luxCDABE promoter–reporter fusion to detect nalidixic acid (NA) as the tested analyte. Light detection is achieved by a single photon avalanche photodiode (SPAD) working in the Geiger mode. Using SPADs allow very sensitive solid state platform operating at signal to noise level superior over simple PN diode based sensors operating without internal gain.

This biosensor can contain several bacteria strains with different gene promoters, generating a unique response signature for different toxins. The system is

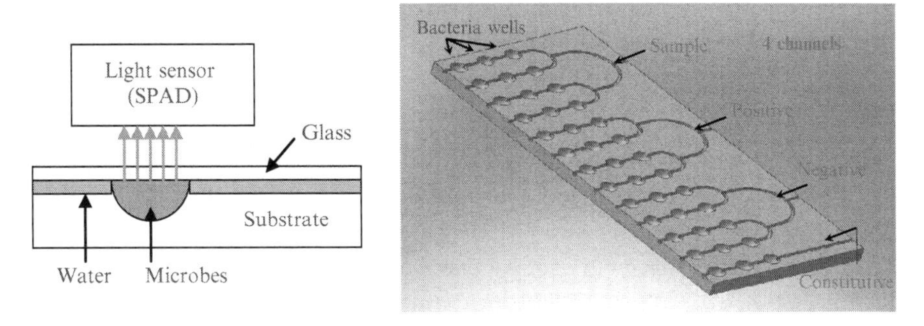

Fig. 5 Bioluminescent whole-cell bio chip: **a**) One cell container and sensor. **b** A 39 cell containers array arranged in four channels: sample under test; containers for testing positive response; containers for testing negative response (drinkable water); constitutive testing of normally bioluminescent microbes

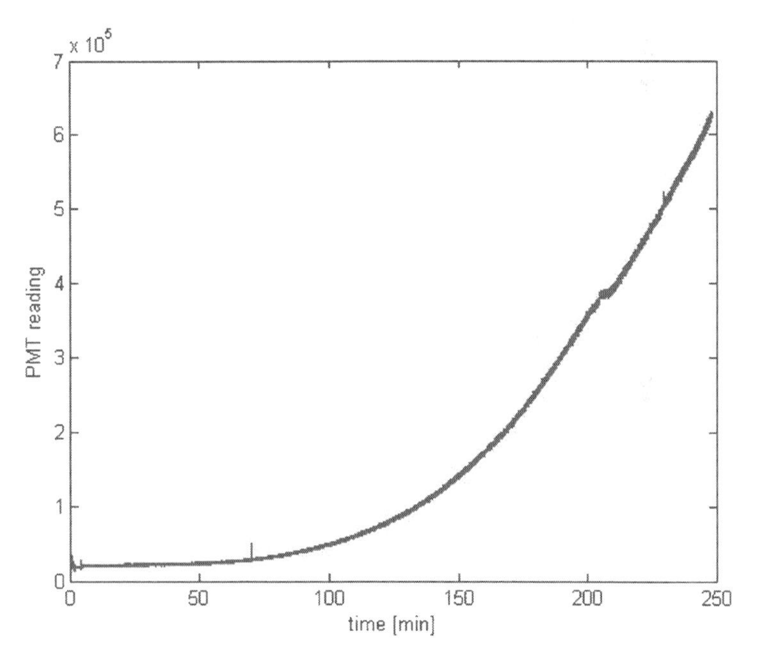

Fig. 6 Demonstration of bacteria bioluminescent response to toxin flow (16 ppm nalidixic acid (NA) with flow rate of 0.22 mL min^{-1}. The PMT reading is shown in arbitrary unit

incorporated into a disposable biochip that provides live cell maintenance and microfluidics channels for sample introduction. The cells are immobilized in the biochip and connected to a microfluidic system which constantly introduces the water-under-test flow into the system Fig. 5.

The biochip consists of the following tests: sample, positive, negative and constitutive. The sample test is the test of the water source. In the positive test, a diluted toxin is injected to the bacteria periodically while in the negative test drinkable water is injected. The sample, positive, and negative tests consist of four strains with a repetition of three bacteria wells for each strain.

The constitutive test consists of one strain which works in "normally On" mode which normally emits lights and stops emitting when exposed to toxins. Measurement of bacteria bioluminescent response to toxin flow is shown in Fig. 6.

6 Summary

This chapter has reviewed different on-chip detection methods for cellular activity. Micro-system-technology enables the reduction of size, thus improving robustness, sensitivity and reducing response time. Different classes of cells can be used in whole-cell biosensors, both eukaryote and prokaryote types. This technology seems to be most suitable for environmental monitoring, health care and medical applications, pharmaceutical and food industry. More applications are expected to emerge as this technology matures.

References

1. Tanaka Y, Sato K, Shimizu T, Yamato M, Okano T, Kitamori T (2007) Biosens Bioelectron 23:449–458
2. Bousse L (1996) Sens Actuators B 34:270
3. Pancrazio JJ, Whelan JP, Borkholder DA, Ma W, Stenger DA (1999) Ann Biomed Eng 27:697
4. Stenger DA, Gross GW, Keefer EW, Shaffer KM, Andreadis JD, Ma W, Pancrazio JJ (2001) Trends Biotechnol 19:304
5. Xiaorong X, Lidstrom ME, Parviz BA (2007) J Microelectromech Syst 16:429
6. D'Souza SF (2001) Biosens Bioelectron 16:337
7. Mulchandani A, Kaneva I, Chen W (1998) Anal Chem 70:5042
8. Reshetilov AN, Lobanov AV, Morozova NO, Gordon SH, Greene RV Leathers TD (1998) Biosens Bioelectron 13:787
9. Ito Y, Yamazaki S, Kano K, Ikeda T (2002) Biosens Bioelectron 17:993
10. Schmidt A, Standfu-Gabisch C, Bilitewski U (1996) Biosens Bioelectron 11:1139
11. Marincs F (2000) Appl Microbiol Biotechnol 53:536
12. Fabricant JD, Chalmers JH, Bradbury MW (1995) Bull Environ Contamin Toxicol 54:90
13. Belkin S (2003) Curr Opin Microbiol 6:206
14. Gilchrist KH, Giovangrandi L, Whittington RH, Kovacs GTA (2005) Biosens Bioelectron 20:1397
15. Borkholder DA, Bao J, Maluf NI, Perl ER, Kovacs GTA (1997) J Neurosci Methods 77:61
16. Chaubey A, Malhotra BD (2002) Biosens Bioelectron 17:441
17. Karube I, Matsuoka H, Suzuki S (1983) Proceedings of the International Meeting on Chemical Sensors, Fukuoka, p 666
18. Reilly A, Kaferstein F (1997) Aquac Res 28:735
19. Yasukawa T, Kondo Y, Uchida I, Matsue T (1998) Chem Lett 27:767
20. Martinoia S, Rosso N, Grattarola M, Lorenzelli L, Margesin B, Zen M (2001) Biosens Bioelectron 16:1043
21. Shibuki K (1990) Neurosci Res 9:69
22. Akaike T, Yoshida M, Miyamoto Y, Sato K, Kohno M, Sasamoto K, Miyazaki K, Ueda S, Maeda H (1993) Biochemistry 32:827

23. Popovtzer R, Neufeld T, Biran N, Ron EZ, Rishpon J, Shacham-Diamand Y (2005) Nano Lett 5:1023
24. Hoeber R (1910) Arch Ges Physiol 133:237
25. Schwan HP, Bothwell TP, Wiercinski FJ (1954) Fed Proc Am Soc Exp Biol 13:15
26. Schwan HP, Bothwell TP (1956) Nature 178:265
27. Guan JG, Miao YQ, Zhang QJ (2004) J Biosci Bioeng 97:219
28. Gomez R, Bashir R, Bhunia AK (2002) Sens Actuators B 86:198
29. Wegener J, Keese CR, Giaever I (2000) Exp Cell Res 259:158
30. Xiao C, Luong JHT (2003) Biotechnol Prog 19:1000
31. Chang BW, Chen CH, Ding SJ, Chen DCH, Chang HC (2005) Sens Actuators B 105:163
32. Chouteau C, Dzyadevych S, Durrieu C, Chovelon JM (2005) Biosens Bioelectron 21:273
33. Abdur Rahman AR, Lo C, Bhansali S (2006) Sens Actuators B 118:115
34. Schwan HP (1963) Physical techniques in biological research, vol VI. Electro-physiological methods. Academic, New York
35. Foster KR, Schwan HP (1989) Crit Rev Biomed Eng 17:25
36. Barber DC, Brown BH (1984) J Phys E Sci Instrum 17:723
37. Metherall P, Barber DC, Smallwood RH, Brown BH (1996) Nature 380:509
38. Bernstein DP (1986) Crit Care Med 14:904
39. Ron A, Singh RR, Fishelson N, Shur I, Socher R, Benayahu D, Shacham-Diamand Y (2008) Biophys Chem 135:59
40. Wang J (2006) Biosens Bioelectron 21:1887
41. Popovtzer R, Neufeld Y, Popovtzer A, Rivkin I, Margalit R, Engel D, Abraham MA, Nudelman R, Rishpon J, Shacham-Diamand Y (2008) Nanomed Nanotechnol Biol Med 4:121
42. Rasooly A, Jacobson J (2006) Biosens Bioelectron 21:1851
43. Sefi Vernick (2008) Private communication
44. Belkin S (2003) Curr Opin Microbiol 6:206
45. Vijayaraghavan R, Islam SK, Zhang M, Ripp S, Caylor S, Bull ND, Moser S, Terry SC, Blalock BJ, Sayler GS (2007) Sens Actuators B 123:922
46. Daniel R, Almog R, Ron A, Belkin S, Shacham-Diamand Y (2008) Biosens Bioelectron 24:888

Index

O

P

R

S

T